# The Joy of Secularism

# The Joy of Secularism

~

## 11 Essays for How We Live Now

*Edited by*

## George Levine

PRINCETON UNIVERSITY PRESS    PRINCETON AND OXFORD

Published by Princeton University Press,
41 William Street, Princeton, New Jersey 08540

In the United Kingdom: Princeton University Press,
6 Oxford Street, Woodstock,
Oxfordshire OX20 1TW

press.princeton.edu

Library of Congress Cataloging-in-Publication Data

The joy of secularism : 11 essays for how we live now / edited
by George Levine.
p. cm.
Includes bibliographical references and index.
ISBN 978-0-691-14910-3 (hardcover : alk. paper)
1. Secularism. I. Levine, George Lewis.
BL2747.8.J69 2011
211'.6—dc22        2010021550

British Library Cataloging-in-Publication Data is available

This book has been composed in Sabon LT Std text with
Centaur MT Std Display

Printed on acid-free paper. ∞

Printed in the United States of America

3  5  7  9  10  8  6  4  2

# Contents

# Contributors

WILLIAM E. CONNOLLY is the Krieger-Eisenhower Professor at Johns Hopkins University, where he teaches political theory. His recent books include *Why I Am Not a Secularist* (1999), *Neuropolitics: Thinking, Culture, Speed* (2002), *Pluralism* (2005), and *Capitalism and Christianity, American Style* (2008). His book *A World of Becoming* is scheduled to be published in 2011.

PAOLO COSTA is a senior researcher at the Fondazione Bruno Kessler (Trent, Italy). He studied at the universities of Milan, Parma, and Toronto. He is the author of two books, *Verso un'ontologia dell'umano* (2001) and *Un'idea di umanità. Etica e natura dopo Darwin* (2007), and of several articles, translations, and collections. He is currently working on the topic of animal intelligence.

FRANS B. M. DE WAAL is a Dutch American biologist and ethologist specialized in the social behavior and social cognition of monkeys, chimpanzees, and bonobos. He is presently C. H. Candler Professor of Psychology at Emory University, and director of the Living Links Center at the Yerkes National Primate Research Center, in Atlanta, Georgia. He is well known for his popular books, starting with *Chimpanzee Politics* (1982), and his writings on the evolutionary origin of morality, such as *Primates and Philosophers* (2006). He is a member of the National Academy of Sciences and the Royal Dutch Academy of Sciences.

PHILIP KITCHER was born in London. He has taught at several American universities and is currently John Dewey Professor of Philosophy at Columbia. He is the author of ten books on topics ranging from the philosophy of mathematics, to the philosophy of biology, the growth of science, the role of science in society, Wagner's *Ring*, and Joyce's *Finnegans Wake*. He has been president of the American Philosophical Association (Pacific Division) and editor in chief of *Philosophy of Science*. A fellow of the American Academy of Arts and Sciences, he was also the first recipient of the Prometheus Prize, awarded by the American Philosophical Association for work in expanding the frontiers of science and philosophy. He has been named a Friend of Darwin by the National Committee on Science Education, and received a Lannan Foundation Notable Book Award for *Living with Darwin*.

GEORGE LEVINE is professor emeritus, Rutgers University, and author of many books on Victorian literature and on the relations between literature and science. Among them are *Darwin and the Novelists* (1988) and *Darwin Loves You* (2006). His recent *Realism, Ethics, and Secularism* was awarded a prize as the best book in the field of literature and science for 2008 by the British Society for Literature and Science.

ADAM PHILLIPS is a psychoanalyst and writer in London.

ROBERT J. RICHARDS is the Morris Fishbein Professor of the History of Science and Medicine at the University of Chicago. He is a professor in the Departments of History, Philosophy, and Psychology. He has authored several books dealing with evolutionary theory: *Darwin and the Emergence of Evolutionary Theories of Mind and Behavior* (1987), *The Meaning of Evolution* (1992), *The Romantic Conception of Life: Science and Philosophy in the Age of Goethe* (2002), and *The Tragic Sense of Life: Ernst Haeckel and the Struggle over Evolutionary Theory* (2008). He is currently at work on a historical and philosophical commentary on Darwin's *Origin of Species*.

BRUCE ROBBINS is Old Dominion Foundation Professor in the Humanities in the Department of English and Comparative Literature at Columbia University. His books include *Upward Mobility and the Common Good* (2007), *Feeling Global: Internationalism in Distress* (1999), *The Servant's Hand: English Fiction from Below* (1986), and *Secular Vocations: Intellectuals, Professionalism, Culture* (1993). He has also taught at Rutgers University, the University of Lausanne, and the University of Geneva, and has held visiting positions at Harvard, Cornell, and New York universities. He serves on the editorial board of the journal *boundary 2*.

REBECCA STOTT is professor of literature at the University of East Anglia at Norwich. She is the author most recently of *Darwin and the Barnacle* (2003), a study of Darwin's barnacle years, and the two historical novels *Ghostwalk* (2007) and *The Coral Thief* (2009). She is currently writing a book of nonfiction for Bloomsbury called *Infidels: In Search of the First Evolutionists*.

CHARLES TAYLOR is emeritus professor of philosophy at McGill University. Among his books are *Sources of the Self* (1989), *Modern Social Imaginaries* (2004), and *A Secular Age* (2007).

DAVID SLOAN WILSON (http://evolution.binghamton.edu/dswilson/) is SUNY Distinguished Professor of Biology and Anthropology at Binghamton University. He applies evolutionary theory to all aspects of humanity in addition to the rest of life, both in his own research and by directing programs designed to reform higher education and public policy formulation. He is known for championing the theory of multilevel selection, which has implications ranging from the origin of life to the nature of religion. His books include *Darwin's Cathedral: Evolution, Religion, and the Nature of Society* (Chicago, 2002) and *Evolution for Everyone: How Darwin's Theory Can Change the Way We Think about Our Lives* (Bantam, 2007). His next book is titled *The Neighborhood Project: Using Evolution to Improve My City, One Block at a Time* (Little, Brown).

# Acknowledgments

Putting together a book of this kind is harder than it might seem, and I owe much to all of the contributors, all distinguished writers and thinkers in their own right, each extraordinarily helpful and encouraging in the enterprise, recognizing the critical need to think again and yet more deeply and compassionately about what it means to be secular, in a world where secularity has not been adequately thought out or felt along the fingertips. I was struck at the outset, when I conceived this book, by how quickly and yet thoughtfully responsive all these very busy people were. Thanks, then, to the contributors, who, whatever the details of the positions they take on these issues, confirmed me in my sense of the importance of the project.

A deep thanks, *grazie mille*, to the Bogliasco Foundation, which gave me the time and the very Italian hospitality that allowed me to write the introduction and gather and edit the contributions. And a warm thanks to all the Bogliasco fellows, but particularly to Kathryn Davis, who, with the wonderful eye and ear of an exceptional novelist, and the acuity of an excellent critic, read an early draft of the introduction; to Eric Zencey, who makes every gelato and every passeggiata a condition for serious and fruitful discussion; to Justin Kramon, whose irony and sense of humor are marks of a first-rate young novelist and a helpfully sympathetic reader; to Orhan Mehmed, an extraordinary musician and scholar of music, whose ear is always to be trusted. I owe perhaps my deepest thanks to Ivana Folle, who, with the most extraordinary efficiency, sensitivity, and personal warmth, helped in almost every stage of my work at Bogliasco.

I want to thank, also, Michael Warner, whose seminar in secularism at the Rutgers Center for Cultural Analysis originally triggered my interest in exploring the complexities of secularism, and made it impossible to remain complacently certain about any aspect of it. Thanks to Gerhard Joseph, who read the manuscript in its early stages, and who remains an intellectual whose fresh and contrarian thought on virtually everything has creative consequences. I want very much to thank my French friend, who helped provide startling information for my introduction, but who has rather nervously asked to remain anonymous. So I thank her here; she knows whom I mean. Thanks, too, to the wonderful editor, Hanne Winarsky, who encouraged me in this scheme, from the time I first suggested it to her, and who has helped thoughtfully and creatively throughout the book's development.

As ever, more than thanks to my family—to Marge, who is always there for me, to Rachel and Dale and their extraordinary children, Aaron and Ben, and to David, whose daughters, Mia and Rosie, are perhaps the best evidence that the world needs no help from the transcendent. Could anything be more beautiful?

# Introduction

## George Levine

Nontheistic reverence for life remains to be cultivated.
—William Connolly (1991)

This book was conceived and largely developed from a totally secular perspective. It will explore the idea that secularism is a positive, not a negative, condition, not a denial of the world of spirit and of religion, but an affirmation of the world we're living in now; that building our world on a foundation of the secular is essential to our contemporary well-being; and that such a world is capable of bringing us to the condition of "fullness" that religion has always promised.[1] The premise of the book as a whole, and probably of all the contributors (but it's important that even that question be left at least slightly ajar), is that the world is to be understood and explained—as far as it is possible to explain it—in natural terms; it works always and everywhere without miracles or supernatural interventions. Despite possible disagreements by some of the contributors, the book is concerned with ideas that follow from this premise, and the central problem is to understand and imagine what it means to occupy a world of this kind. Through explorations of this problem, I want to establish an affectively and intellectually powerful case that a secular world is not only worth it—that life is indeed worth living—but that with all the inevitable pains and losses, it can be wonderful, indeed, at times joyous, and that a sense of that wonder enhances the possibilities of improving it.

Philip Kitcher, a distinguished philosopher of science, sets the tone for the book, opening the discussion with a secular vision, confronting the challenges to it, but finally arguing that "an adequate response to these challenges requires moving beyond secularism as a merely negative doctrine, and offering something to replace the functional aspects of traditional religions." "Functional" may sound to nonphilosophical ears a little too businesslike for the achievement of "fullness," but in replacing the "functional aspects" of religion, one needs to achieve as well the kinds of emotional, personal, and cultural satisfactions that religion has traditionally been conceived to provide. It's for this reason that I (like several of the contributors) have hung on to the rather problematic but often extremely useful word "enchantment." If we take that word carefully—with the qualifications implied by the rest of this introduction and perhaps with some of the reservations of Bruce Robbins in his essay here—there are ways in which the secular world can be experienced as "enchanted" and remain absolutely of this time, of this place.

A book about ideas, about science, ethics, history, and art, this volume nevertheless aims also at being a book about living: how does it *feel* or how *might it feel* to be entirely secular? how can we, as Paolo Costa asks, feel ourselves at home in a world that allows for nothing transcendent? how can one come at things, understand them on their earthly terms, and think about them, experience them as value laden and "meaningful"? I have undertaken this volume to explore these questions because it is a matter not *only* of ideas, but also of experience, and because by learning to be at home in this secular world, we can feel more intensely the urgency and understand more fully the possibilities of making it better.

There is, then, real practical urgency in what, to some, might seem rather arcane struggles over a theory of a sociologist, Max Weber, who died eighty-nine years ago: "the fate of our times is characterized by rationalization and intellectualization and, above all, by the 'disenchantment of the world.'"[2] Science will figure importantly in the arguments that follow because it is our best instrument for discovering how the world works; and knowing as much as we can about how the world works is a condition for improving as much as possible our human condition. Yet the Weberian story is that science "disenchants" the world, takes the "meaning" out of it—and it does so not so much by fully explaining it, but by making it seem as though it *might* be fully explained without recourse to the invocation of transcendental powers. But while Copernicus's famed decentering of the world and thus of the human was the beginning of a long history of decentering that has played itself out in physics, in many important psychological theories, and in Darwin's theory and the evolutionary biology that grew from it, the argument here—all the contributors seem to agree—will be that the news science brings is not in the end anything like the spiritually devastating news that Weber's story of disenchantment seems to imply.

Behind these arguments about "knowing" and "feeling" and "ethics" and "enchantment" is a conviction that secularity is not an option but a necessity, an absolute condition for a democratic society and a democratic world. Competing forces, competing religions, make claims for ultimate authority not only over the life of the spirit, but over the conduct of our lives in this world. The moral and spiritual authority that religion inevitably claims *must* imply universality, and this has to mean in all consistency that each religion must demand moral and spiritual control over lay society as well. Clearly, many religions, or many religious people, have backed off the universalizing claims that religion as a whole seems to imply, and agree to respect alternative ways of believing. But the examples of conflicting claims are everywhere, and for survival—and, certainly, for any condition richer than mere survival—nations internally and in their relations with others must allow some authority *other than* the factious ones that competing religions represent. Nobody can sanely assume that secular societies are, in contrast, always cozy and harmonious (a little Stalinism, a little Hitlerism, a little Maoism will blast that absurd notion out of the water), but it makes sense to believe that for a democratic civilization to survive, there must be some authority granted (by all the contenders for the "true" truth and the ethical best) that is enfranchised to limit their power to impose on others. Whether that is ultimately possible and sustainable is an entirely unsettled question. But that ultimate authority—the democratic state—must not be defined as belonging to any of the contenders.[3]

On this account, a truly democratic state can't be Christian (or any of the Christian subsets or alternatives, like Roman Catholic, or Jewish, or Islamic), but must be ready to allow Christianity, or Judaism, or Islam, its space within the community. There must be what Charles Taylor calls a "theory of social order . . . that transforms our social imaginary."[4] Which is to say that a truly democratic state must be secular, whatever any individual inside the state believes about the transcendent and its relation to our ordinary lives.

This book, then, with whatever individual qualifications any of the contributors might make, is after something like what Taylor describes in his essay "Modes of Secularism."[5] There, Taylor—a believing Roman Catholic—sees secularism as a kind of moral imperative in the modern world, one that allows for the extraordinary range of beliefs among people and that makes democracy at least possible. What is required is a social and political organization that is not governed by any one of those myriad beliefs but is willing to acquiesce, in order to make civilized life possible, in an "overlapping consensus" on culturewide decisions about the particulars of ordinary and social life, on the law, and on the rules of social behavior.[6]

In order to arrive at such a development devoutly (well, yes, devoutly) to be wished, it is necessary to reimagine secularism, recognize in it all of the ethical, spiritually satisfying, and deeply felt (not merely dispassionately held) elements that have traditionally been allied with religion, but without the magic, and without the sectarianism. A tall task, perhaps an unworkable one. But it is a task more than worth the effort. To that end I invited a broad range of thinkers, among the most distinguished on such subjects in the world, and I asked them, "Would you think about the possibility of 'Secular Enchantment'"?—that is, about a "positive" secularity that might satisfy those crucially important feelings and spiritual needs that it would be self-destructive and self-deceptive for secularists simply to deny. Obviously, I asked writers whom I thought of as more or less secularist in orientation but, most importantly, writers who had clearly thought hard about these problems. I did not know what they would want to say, only that what they would say had to be considered seriously. My own object, whatever the authors themselves would decide, was to disentangle morality and "fullness" from the transcendent, to shift attention and emotional investment from the *aldilà* to the here and now. Any large movement toward a democratic secularity requires that we beat that "disenchantment" story and allow ourselves to feel deep in our bones the compatibility between profound attention to the details of this world (which are so often for so many so preponderantly painful) and a sense not only of material but of what we call spiritual satisfaction.

It will be no surprise, then, that I invited William Connolly to write for the book, since in an early essay, which in fact provides the epigraph here, he had made a powerful case for such a project. In his "Letter to St. Augustine," Connolly movingly engages with Augustine, considers his arguments, and complains: "The more one looks into the depths of the human condition—and you peer more deeply into this abyss than anyone before and most after—the more it becomes clear how much must be divested from human life and invested in divinity if the very possibility of salvation by a sovereign god is itself to be made secure."[7] This book's aspiration to imagine a positive and satisfying secularism is directed against such a divestment. Isn't there something of a paradox implicit in the idea that in order to achieve satisfaction of the kind that Taylor describes as "fullness" (or the salvation Augustine desires) we need to seek explanation outside the world that inspires the satisfaction in the first place, and detach ourselves from our own world, devalue it, or even deny it, turning what common sense suggests is reality into a mere illusory passage into the real reality? That, at least, is what I was thinking when I undertook to put this book together. The writers on the whole have independently worked out approaches to the problem that I have found compatible with my initiating idea, although often complexly different as well.

Together, they seem to me to provide a rich and exciting set of arguments about the possibility and necessity of imagining a positive, if often dangerous, secularism and finally living full lives within it. And for the rest of the introduction I will be talking about this possibility with rather less philosophic precision and rather more sentiment than do several of the contributors, trying to read the contributors' essays as they reflect on the possibility of such secularism, while trying also to suggest in my own terms what it might mean to live a satisfying secular life.

ↄ⁄ↄ

Darwin once wrote that he wondered sometimes why everyone didn't want to become an ornithologist. Well, I do. I love to watch birds. I love birds. And I continue to be astonished by their variety, their intelligence, their beauty, their (alas, decreasing) ubiquity, and, in all that, their difference from me. Somehow they manage in the depth of urban jungles as in jungles proper. Rain forests pulsate with them and reverberate with their calls. The Ramble in Central Park fills with them in the spring. But the streets of New York (and Los Angeles, and Chicago) also sustain them—pigeons, house sparrows, starlings, house finches, mallards, crows, red-tailed hawks, herring gulls, now again peregrine falcons, and even communities of descendants of escaped parrots. I might even say it's a kind of miracle, this vitality and adaptiveness and variety. All one has to do is look and there they are. We imagine some great distance between our urban selves and nature, but the birds remind us that we're part of each other, depend on each other.

Thinking what the experience means can belie the intensity of the perception and response. It's better for me at first to say "wow" than to puzzle about where they come from, how they sustain themselves, how they resist the murderous or indifferent presence of millions of humans and their machines. It's better for me to notice what their commonness can easily disguise, that these birds are beautifully feathered, prolific, graceful, quick, and clever and find the resources they need in garbage, crumbs, manure, the leavings of human excess. They get so common one sometimes wishes them away, but who has not been awed by the astonishing performance of those pesky, inextinguishable starlings as they swoop in clouds synchronized but irregular.

And the feeling, the deep, let's call it "natural," satisfaction of recognizing so many varieties of life independent of us and yet so clearly related, feeds that sense of "fullness" that gives to our lives a "richness," makes it feel "deeper, more worth while, more admirable, more what it should be."[8] For Taylor, such moments point upward, toward something outside this secular world that provoked it. For me it has been different. It points me—let me follow the metaphor—downward, or up only so far as the birds fly. This world seems to me, amid the urban noises, the garbage, and the disasters that announce themselves regularly in the papers and on TV, very beautiful and adequate to itself. The earth is room enough—or should be. So that those moments of fullness with which my life has been so fortunately enriched are deeply, even passionately, secular moments. "Wow." What a world!

And yet of course, in the long curve of the experience of Western culture, and perhaps of all cultures (I'm no expert), there has pervaded a sense that the very value I experience watching those pigeons and hawks and ducks, and the ethical and aesthetic values that give our lives fullness and sustain our societies and our culture, are somehow indissolubly involved in the experience of another world, the *aldilà*, the world that religion represents for so many cultures. Against the secular imagination of the world that I feel to my fingertips, there emerged right from the start, through the very heart, for instance, of nineteenth-century culture where secular talk was getting so much greater play, intense counterarguments: the world without religion couldn't account for it, and was simply dog-eat-dog, amoral, violent, and without sanction for the ethical. We are familiar with the powerful Dostoyevskian sense that there is no meaning in a world without God. Similarly if less powerfully imagined, this view held among most Victorians. W. H. Mallock published a book called *Is Life Worth Living?* (1879), and, echoing that, William James published an essay of the same title (1896). It was a real question: if one removes God from the world, can there be any meaning, any value, any morality? Is there any point in being alive in such a world, just one damned thing after another? (Perhaps the word "damned" here has more force than is usually registered in the colloquialism.) For Mallock, the answer was no; and, in a different register, and in a more complicated way, so it was for James. And as I watch, with awe and delight, the red-tailed hawk swoop down to the apartment house just across the street from Central Park, I occasionally ask myself, why should that be so? Isn't this, right here, worth it, after all?

It's a serious question, obviously. The answer I propose by putting together this book is "of course," and it is driven by a powerful desire to make that "of course" simple common sense for as many people as possible. At first, I want to insist, one doesn't have to think about it. There's the bird. Not that all the contributors agree with me, or even consider the question immediately in such absolute terms; but I want to link together that immediate almost mindless experience of the wonder of the ordinary (at the same time, profoundly Darwinian)[9] with the larger question "Is Life Worth Living?" which I might translate into: can the world perceived as entirely secular, explicable exclusively in naturalist terms, with no recourse to what Daniel Dennett contemptuously calls "sky hooks"—can it provide the kinds of "fullness," the kinds of moral, aesthetic, and spiritual satisfactions that have traditionally been linked exclusively to religion? "Of course" may seem too easy, and the writers for this volume explore this idea from many vantage points, sometimes with great complexity and sometimes with implicit disagreement, but wherever they end up going, I—as editor with only personal authority—want to insist on the "of course," on the fundamental ground of feeling for secular experience of value.

The problems with my happily middle-class and intellectual enthusiasm leap immediately to mind. James's essay begins by invoking a joke he says made the rounds after Mallock's book asked the question "Is Life Worth Living?"—"it all depends on the liver." The joke is, in the end, no joke at all. It would be absurd for an academic bird-watcher, in the midst of a fairly healthy and comfortable life, to claim for a world full of madness, suffering, starvation, predation, Ponzi schemes, drug cartels, disease, earthquakes, volcanic eruptions, floods, and genocides to claim, with the complacency of good luck, that everyone should just sit back and smile at how wonderful the world is—just appreciate it! James wonderfully insists that there are some people—he invokes, for example, Walt Whitman—who have "instinctive springs of vitality" and are unreflectingly and temperamentally upbeat, people for whom the world can be wonderful, the blade of grass a miracle to stagger sextillions of infidels. But there are others, perhaps with deficient livers, for whom, even if their lives *are* otherwise comfortable, the temperament points downward to darkness. "We of the nineteenth century," says James, "with our evolutionary theories and our mechanical philosophies, already know too impartially and too well to worship unreservedly any God of whose character [nature] can be seen as an adequate expression."[10]

In this volume, Adam Phillips discusses the Freudian perception that we are all born into "helplessness," a permanent human condition. We come into the world utterly dependent, and the helplessness does not abate but transforms as we grow. The condition ensures that the world is not likely to be a lot of fun, not, certainly, for long; it is overpowering and scary; and humans simply don't have the resources to resist its forces. We are helpless. Facing that helplessness, on Phillips's account of Freud, is what we all must do, short, I suppose, of the suicide that both Mallock and James discuss as the logical response to the godless human position. Ironically, as Phillips sees it, the attempt to find a helper—and the most potent final helper would be God—is a way to disguise the reality that James, too, had confronted. James's answer, in the plea for "help" that his essay might be taken to be, was to translate longing for God into the reality of God. (And this leads us, as do several of the essays in this volume— particularly David Sloan Wilson's—to a question not only about what constitutes the truth in our descriptions of this world, but whether truth, in our rationalist/scientific mode, is what will help us best.) Freud sees such a move as a surrender, a retreat from reality, and, perhaps ironically, a disguise of real (I might say, secular) desire. And, Phillips's essay implies, it is only satisfaction of that earthly, secular desire, which can come in moments as we live in the here and now, that in the end makes life worth living. So the best thing to do, ironically enough, is to surrender to the inescapable fact of our helplessness, and find then the sporadic satisfactions of worldly desire, lost in the help-seeking flight to God.

It gets complicated. There are lots of ways to read the conditions of secularity, and when I invoke Taylor's idea of "fullness," it is in part be- cause I understand "fullness" to be not an unequivocally happy state, but a condition, sometimes painful or worse, that allows us nevertheless an ultimate sense of the value of things and of ourselves. So in this book I want to tangle with the complications and preserve my "wow" after all. Can we allow that we are "helpless" and yet experience "fullness" in the secular condition? "Of course."

Although I am deeply tempted by a rhetoric of heroic facing of the truth and have been in sympathy with the "freethinkers" who in the late nineteenth and early twentieth centuries largely invented it— writers like T. H. Huxley and Karl Pearson and Leslie Stephen and Ber- trand Russell—I realize that there are problems with its ultimate dramatic complacency, and perhaps even with its epistemological assumptions. Holding tight to inconvenient truths, we somehow portray ourselves as intellectually and morally superior to the rest of the "big children"—Max

Weber's words[11]—who imagine that the truths of science are compatible with religious faith. On this side, there is the inescapable if constantly self-correcting truth of science, teaching us how to maneuver in this world; on the other, there is religion—teaching us how to maneuver in the other world, and, as a route to that other world, policing us in our behavior in this one.

Unlike others, like the late Stephen Jay Gould and so many excellent scientists who believe that there is nothing incompatible between science and religion—nonoverlapping magisteria, Gould somewhat bloatedly calls them—this book will make no such case. Eschewing the triumphalist rhetoric of brilliant writers like Richard Dawkins and Daniel Dennett, I nevertheless accept by and large Weber's, shall we call it, somewhat snotty division between the two. The point of this book is not to try to reconcile science and religion—some people might be able to bring it off, others certainly not, and I don't really believe in it; it is, rather, to imagine a satisfying secularity, a secularity with which "fullness"—not traditional religion—is compatible.

The rhetorical problems, meanwhile, are substantive ones as well, for the argument from a strict scientific rationality assumes that once, within the epistemological framework of the modern West, one finds "evidence" and the logical relations among ideas and things, the case is closed, and this despite the fact that nobody except the previously unconverted is persuaded by the unexceptionable reasoning. The rhetoric that follows tends, I fear, to be triumphalist and pretty intolerant of anyone who either denies or questions the "evidence," or anyone for whom the assumptions about the priority of "reason" over "faith" are in question. Yet there is a strong philosophical case to be made that scientific reason, too, is based on a kind of faith of the sort that Hume discussed centuries ago.[12] Certainly, the deep modern confidence in science's absolute authority has not always succeeded in achieving the kind of intellectual or moral effect toward which it has always been directed: those who "cannot bear the fate of the times like a man," as Weber put it,[13] make a terrible "intellectual sacrifice" in their refusal to accept inconvenient truths. I confess to having bought in pretty thoroughly to this way of thinking, and in my recent anger and dismay at fundamentalist responses to fundamentalist attacks, I have found it a deliciously tempting option to join militant atheist rhetoricians.

But one way to get at what I am after in this book without mounting a moral and intellectual high horse was articulated in a recent *New Yorker* article by James Woods. "What is needed," he writes, "is neither the overweening rationalist atheism of a Dawkins nor the rarefied religious belief of an Eagleton, but a theologically engaged atheism that resembles disappointed belief. Such atheism, only a semitone from faith, would be, like musical dissonance, the more acute for its proximity. It would give a brother's account of belief, rather than treat it as some unwanted impoverished relative."[14] But the problem is not only a matter of avoiding triumphalist rhetoric while at the same time avoiding belief in the improbable. Woods's excellent and sensitive formulation falls just a semitone short of what I am after.

From my perspective, in seeking an engaged atheism, Woods puts slightly too much pressure on the *disappointed* nature of the belief. It is easy to recognize where that comes from, although I don't think that I personally have ever been disappointed in that way. I want, rather, as the old song suggests, to "accentuate the positive." The story of secularism, even (or particularly) the brilliant and almost exhaustive one told by Charles Taylor, is for my taste and beliefs too ridden with "nostalgia." And that, of course, means homesickness. How lovely it was to believe! Home is where we were, not where we are! And thus the disappointment implies that we want to go home again but can't: just exactly, for my purpose, the wrong way to look at the world. This *is* our home. So all of this kind of thinking and feeling makes a version of the narrative of disenchantment that any discussion of secularism these days must confront. It was better in the old days when there was magic everywhere, and where religious and secular authority were joined—in some rather brutal structures of social hierarchy. If nostalgia for a lost and lovely religion in which we can no longer believe leaves us forlorn, then I choose here to refuse nostalgia and the sense of loss, and I want this book, if with sympathy for those who feel the loss, to take me in another direction.

Woods's formulation admirably entails a respectful nostalgia, but Philip Kitcher, in his essay here as in his recent book, *Living with Darwin*, gets to where I think we should be going.[15] Against the triumphalist rationalist rhetoric, as against the "intellectual sacrifice," Kitcher recognizes that secularism requires something other than the denial of the transcendent, and something better than the lament for lost improbability. I quote again his formulation of the crucial project: "An adequate response . . . requires moving beyond secularism as a merely negative doctrine, and offering something to replace the functional aspects of traditional religions. Secularism needs to become secular *human*ism."

All of which suggests that there is indeed a way, as fundamentalist critics of secularism angrily insist, that secularism is *only* a kind of alternative religion—but a worse one, and of course, as such, it is subject to the same critiques it levels at religion itself. Even if one is determined not to allow "secularism" to be understood as a negative of religion (as atheism is properly defined as the negative of belief in God), there seems to be no way to escape the shadow of history, and therefore of religion, which emerges from the deep and long continuing past. Since it is normal in our history to associate religion with the quest for meaning and the possibilities of fullness, secularism as (and if) it grows will certainly owe something—as Taylor argues in *A Secular Age*—to religious traditions, and thus Woods's formulation, accepting that, moves in the right direction. Offering something to replace the "functional aspects of traditional religions," including, I would add, the experience of "fullness" with which this introduction began, entails an inevitable even if unwilling paralleling. In his essay, Kitcher claims that to produce an adequate secularism there must be a detailed understanding of what it is that religion supplies. And we have already a significant tradition of secular study of religion in this direction, the work of Durkheim, for example, and the more recent work of David Sloan Wilson,[16] that emphasizes, among other things, the indispensable role religion has played in the binding of communities.

But this absolutely must not mean "religion" as we are used to thinking about it; it must mean *displacing* traditional religion from areas properly and significantly belonging to the secular world. It must not mean, to take the most extreme example, producing the imitation religion of nineteenth-century Comtean Positivism, with its chapels, churches, breviaries, and the like. Facing in an earlier phase the intellectual and spiritual crises with which this book is engaged, the Positivists, in some ways quite sensibly, made up a church and worked up a list of rationalist saints. The absurdity of the effort was clear almost from the outset—what we do *not* need are artificial rituals, imitation catechisms, formal Sunday meetings and "lay sermons," rationalist confessionals, and the rest.[17] All of these are confessions of loss and in imitation are inevitably weak and even comic versions of a historically dense and impressive tradition. What we do need is an understanding that the most important provinces claimed by religion—the provinces of ethics and art and daily life—are provinces that secularity has a powerful right to reclaim. Kitcher and Woods share a tendency toward turning secularism into an alternative religion because they understand—as does Connolly in his superb book, *Why I Am Not a Secularist* (though he is)—that rationality constitutes only a small part of the human experience and is helpless against the urgencies of feeling. The important thing is to find a way to conjoin the worlds of feeling with the worlds of reason, and one must do it with another kind of rhetoric (rhetoric itself being defined as just such a combination), and another way of living. What we seek, as David Sloan Wilson puts it here, is "a cultural system that delivers the best of what religion has to offer while respecting factual reality." Yes, as Kitcher says, "secular humanism": but not a religion, rather a way of being that occupies ground too often claimed as exclusively the province of religion.[18]

The point—and it's perhaps the unreflective point from which I started—is not that secularism is a falling away, a necessary loss—but that (and I'll sneak that word in again) it's a damned good thing. As Bruce Robbins wants to insist in his essay here (in which he takes on the whole tradition of what he certainly thinks of as false and obscuring nostalgia), secularism *should* be seen as "an improvement story." To suggest something like progress, in the current intellectual culture that refuses the idea of progress even to Darwin, is actually quite daring, although scientific/rationalist thinkers will have little trouble with it. To resist the idea that the world gets better and better and will eventually achieve through technology and human inventiveness a kind of utopian future seems to me to be crucially important. Buying into that version of the progress story means buying into a whole bunch of political and economic narratives that need to be challenged; it turns the "reenchantment" story into a *secretively* political program, decried on the one hand by Eagleton, who leans against it back toward the traditional religious position, and on the other by Robbins in this volume, who does more than lean away both from the enchantment narrative *and* from the traditional religious position.

But, even in the light of many intellectuals' deep disappointment with the exploitation of scientific ideals of progress in the development of vast industrial, technological, and financial systems that have moved Western culture to the edge of catastrophe, it makes important sense to recognize developments in scientific thinking as at least epistemological "progress." (And that progress, as I will be arguing, can be linked crucially to ethical "progress" as well, though the matter is much more tricky and difficult there.) We do know more about how the world works; we are busy, self-consciously, finding out how much more we can know and whether material processes can serve as explanation and then as help (it's dangerous but irresistible) in what we have traditionally taken as the province of the spirit and of the deeply private spaces of our selves. The problem of "reductionism" rears its head here: can we reduce everything to some fundamental material condition? Certainly, what I would call premature "materialism," as I think of it, produces some awful stuff: books that prove that men are naturally rapists, pills that "cure" everything from homosexuality to spiritual darkness. But I still want to know, as Frans de Waal tries to teach us, something about the potentiality for morality in primates; I want to know, as David Sloan Wilson tries to discover, how much aspects of human behavior, like our rituals and attitudes, have almost unbeknownst to us served adaptive causes. I want to know if there are elements in the brain that are automatically, physiologically "sympathetic." I want to know a lot, and to be free to investigate "scientifically" spaces previously closed off by our notion of the sacred. "There is nothing specifically religious about the concept of the sacred," says Wilson.[19] Knowing more about the way things are is a condition, as almost all the essays in this book imply or argue, for doing anything valuable to them or with them.

At the end of her essay for this volume, with an appropriateness I could only have wished for, Rebecca Stott cites Amy Clampitt's poem, "The Sun Underfoot Among the Sundews." Its last extraordinary image captures richly the conception of secularism I am trying to get at all too prosaically: "There is so much light / in that cup that, looking, / you start to fall upward." The echo of religious language—"the fall," most obviously, and the "light," traditionally its antithesis—leaves the world we occupy upside down. Stott's analysis gets it right, Darwinianly right. In a secular world the "fall," which is literal, displaces the upward movement of salvation in the religious world, and puts us among very beautiful things and among dangerous ones. One falls upward into the gorgeous predatory, insect-devouring sundews. While they are gorgeous, it is impossible to forget that those flowers are "devourers," that they exist in a bog, and that they provide no consolation but their own dangerous beauty.

This is where the secularism I am imagining for this book will take you, with the hope that the "fall" will be recognized as upward, rather in the way, although in so different a form, the accession to "helplessness" in Phillips's essay is paradoxically an invitation to the morally and aesthetically satisfying life—to the possibility of "fullness." Simple atheism is, Philip Kitcher reminds us, a fundamentally negative position—it is a "denial" of the existence of something, God, for which it cannot find, within its own canons of proof, any evidence. But such epistemological (and metaphysical) assertions are beside the point of our current religious/secular stalemates, and often deadly wars. Following Clampitt's poem into the brilliant bog, we come to the position—metaphorically speaking, of course—of treating secularism as Kitcher argues it must be treated if it is to have any purchase on the real life of our time (or any time). That is, it must be articulated "as a set of positive responses" to the losses incurred when one surrenders the positive ways religion enters people's lives—through felt inspirations to generosity and moral behavior and to the building of community, for example. Somehow, that is to say, secularism must make not the epistemological but the social and cultural appeal that religion has long been understood to make. Falling up into the sundews is what we want to be able to do. Much of the energy for the writing of this book comes from a deep faith—yes, the word is necessary, though I think there are tons of evidence to help justify it—that secularism is a "falling upward," and that such secular appeals are possible.

Taylor, Robbins, and Robert J. Richards remind us that Max Weber's word, normally translated as "disenchantment," is actually *Entzauberüng*, the removal of magic; and the move from magic to science is critical in what amounts to a struggle between traditional improbable if lovely stories and modern probable ones to determine, as Weber did, that the probable ones are better (I want to insist that they can be lovely, too, though that hard sell is part of the work of this introduction, and of the book as a whole). And they are better not only because they are more likely to be true. The whole question of the difference between truths that are useful—or adaptive or pragmatic—and those that are objectively or scientifically true is central to the question of secularity and is at least partly treated here in Wilson's essay. But the bias of this book, or at least of this introduction, is that "true" truths as opposed to adaptive or pragmatic truths are likely, in the end, also to be more adaptive and pragmatic too. It is humanly better, suggests Phillips, to acquiesce in the reality of our helplessness. It is ethically more efficient to find out whether, as some have claimed, altruism is incompatible with our lives as organisms governed by the processes of natural selection than to assume a transcendental insertion of altruism into our nasty secular bones. The narrative of secularism I mean to imply in this book is, then, an improvement narrative, one that winds through the extraordinary difficulties and complexities of humans' understanding of what it means for a thing to be true to an understanding of truth that allows the inconvenient truths their space and opens the way, nevertheless, to the satisfactions that religion was the only institution to supply when such truths were scarcer.

It sounds, even as I write it, a bit callow and presumptuous, but think of it as a fall upward into the sundew. There it is. Secularism is to be understood or felt not as a necessary loss, a heroic counterstory to a religion that couldn't, alas, fulfill its promises, but as a recognition of the brilliant sunlight down in the bog, and thus a profoundly positive affirmation. Yes here. Yes now.

The enchantment narrative against which Robbins writes with such intensity does tend to imply the superiority of the "then," even if we know that the "then" has to be left behind. It implies, as does so much of Taylor's wonderfully comprehensive *A Secular Age*, even with its remarkable understanding of the spiritual attractions and possibilities of secularism, a nostalgia for the lost enchantment, the lost magic. It is a long and moving story, but, as I am arguing here, it tends, however wisely, to turn the magical world into home. Against that imagination, or perhaps it would be better to call it implicit feeling, one might consider the conditions possible in a magical world.

I recall some descriptions of a culture in the very heart of Central Africa as a French friend experienced it only a few years ago. She spent a long time living among the natives of that area, and entirely on her own. In her early weeks there she was fully accepted by the community and felt no fear. She had long conversations with the witch doctor, was befriended by men and women and children, and learned as much as she could about the culture of the community. She reported to me, however, that the natives— like other peoples living all around them—believe in a mind-full, intentional world, a world that Daniel Dennett might say is built on "sky hooks." Everything that happens in that world is "intentional." There is a mind at work always, and what we call magic they take as the quite reasonable norm of everyday life. What for us is irrational is for them precisely the reverse; and I'm sure vice versa. Among those peoples there are pretty heavy penalties when the "mind" behind the accident is identified. An accident doesn't happen. Someone is responsible and must pay.

So there are no accidents. There is a reason for everything. One may not see it, but mind is everywhere at work, and these natives seem *always* to have a feeling the equivalent of what many people in Western cultures often claim when they survive a crash in which most other people have been killed and wounded: "There must have been a reason I was spared. Thank God for saving me" (which of course implies that there was a good reason that all of the others should be dead).

It is, then, a world guided by magic. Magic makes the common sense. If someone stubs his toe or falls or dies, it is important to find the person responsible. The witch doctor finally determines who is the culprit (not *whether* there is in fact a culprit), and consequently there are more than often enough quite terrible punishments meted out. My friend claimed that she had seen the very gentle and friendly people she had come to know execute violent punishment, even kill, because someone had drowned or had fallen accidentally to his death. Someone was at fault, was guilty, and had to pay. The decision was absolute, and its seriousness was played out in the degrees of punishment inflicted. "La mort est partout," my friend said. When she herself ran into some little trouble because of her encounter with another witch doctor, her comfort with the lovely people turned into a frightening sense of vulnerability, and she pretty much stopped going out. Her life, too, might have been in danger, and one of the witch doctors might easily have rendered a negative judgment about her. The last relevant part of this story is my friend's conviction that magic and witch doctors would almost certainly have disappeared from the culture had there been a decent hospital.

In my perhaps hopelessly Western reading, the culture my friend entered was "enchanted," was full of magic, and made complete sense while in a space of total mystery. That is, while things made sense because the world was governed by mind, what exactly that mind was thinking or would do remained mysterious. It is almost the reverse of Westerners' view of the world: natural selection is mindless and its processes are without moral responsibility. We may be horrified by the destruction, but we understand what Dennett would call the "algorithm" of its working. The mindlessness science identifies opens up the possibilities of understanding. The world of natural selection is "disenchanted," but the consequence is that we know more about the world than do those who believe that all life is governed by mind and intention. The very rational, mind-filled existence of these African natives produces just the conditions implied by "enchantment."

The religious worlds of the West enact a similar story: terrible things happen, but surely they are part of God's plan. In an enchanted world, as Weber emphasizes, there is always mystery, something one doesn't understand working importantly through life, but the culture my friend described to me is one in which there is meaning saturating all things in nature even if it works like magic and is in most respects a mystery. The ultimate reasons remain a mystery, but we know there is a mind that makes sense of them—a mind out of nature. The joy of enchantment—like Don Quixote's translation of all he sees into intentional, meaningful action—is just that it provides the satisfaction of some ultimate moral and even rational frame (even if we are too small and humble to know what it is). This kind of mind-filled world is, in its detailed processes, magical. *Mais la mort est partout.*

Looked at from this perspective, enchantment loses a lot of its charm and nostalgia comes less easily. As it did my friend, it can leave you right in the center of the magic, but unpredictably vulnerable and not at home at all. From my perhaps shallowly secular perspective the "mindless" world of natural selection seems a considerable and much more comprehensible improvement: we can understand the processes of natural selection without feeling the necessity of punishing it (or any of its henchmen). What has happened in that change from magic to natural processes is something like what Taylor describes in his essay here.

On Taylor's account, then, secularity is marked by the fact that—to borrow a formulation from a novelist rather than a philosopher—all the action is thrown inside. Value exists not in the things, in the stones or in my birds, but in "what we call minds," and "the only minds in the cosmos are those of humans."[20] In the "enchanted" world, matter was inhabited by spirit; value existed outside my perception of it. Darwin's imagination, as Dennett often all too gleefully puts it, is of a world initiated out of mindlessness and developed mindlessly.[21]

But if I want to think of this view as an improvement, I must also come to terms with the problem Taylor presents: when I want, from the "disenchanted" secular condition, to make the argument that the world *is* wonderful, or beautiful, or to argue for some ethical imperative, I am in a pickle. If I want to claim back for *this* world, as I do, those qualities of fullness and richness that, supposedly, were present in the enchanted world, say, of the pre-Renaissance West, or of my friend's Africa, I will need to make those value arguments stick. It is not that the world is beautiful just for me, but that it is, simply, beautiful—even, say, mosquitoes. Following this kind of argument, in order to find meaning in the world, I would somehow have to make universal claims.

I have to believe, as Taylor so carefully puts it, that a "strong evaluation" implicit in this kind of claim will not allow us to release ourselves from it; even for secularists who deny external sanctions, any attempt to do so—say, to violate one of the commandments or concede the relativity of value (how about "Thou shalt not kill"?)—"reflects negatively on us." And even my sense of wonder at that stooping falcon, diving at speeds that make its prey quite literally explode on contact, is a "strong evaluation." It *is* wonderful (and that includes the destruction of the pigeon, another being). It is for me, as Taylor says, "the truth of the matter," and that, given the perspectival relativism still latent in such strong judgments, contradicts itself. This unsteadiness of meaning is often countered, in current discussions of ethics and morality and religion (as in the Durkheimian position espoused by Wilson about the adaptiveness of religious belief) by a move to an extratranscendental explanation of matters ethical, in biology. De Waal shows us here that some primates tend toward altruism (at least in certain situations). In another register, Richards, with the most careful textual and historical analysis, shows that the determinedly naturalistic Darwin "provides the grounds . . . for an ethics that meets the standards for a normative system." That is, there are strong arguments (much contested, of course) that one can move from naturalism to ethics, or at least find the grounds for strong moral judgments in natural processes.

But here Taylor points to another problem in my little battle to insist on the possibility of fullness and ethical urgency inside the secular. That is, that what we have meant by altruism in the long history of Western and Christian culture exceeds the biological explanation—there is a sense that "admiring altruism is not just finding this pattern of action useful [or, in perhaps more sophisticated evolutionary terms, serving without any selfish motivation but merely instinctively, to be useful to the group]; it is also holding that the motivation which powers it is in some ways higher, more noble, more admirable." Taylor allows secularism to do a lot of work, but he digs in his heels at a position that offers quite mundane and pragmatic explanations of spiritual conditions of the sort, you-scratch-my-back-and-I'll-scratch-yours.

I want to agree with Taylor's implicit point, that a full secularism leaves us in a very weak position for formulating a universal value imperative. I want, however, tentatively though it must be, but in the company of others of the contributors to this volume, to settle for something less than the "strong" imperative, and to recognize Darwinianly (as certainly Taylor does far better than I) that there are gradings and differentiations, and to turn my universal imperative into a demonstration of possibility. It is not as though those who believe that ethical assertions are universally valid actually, in practice, agree about what the ethical universals are. I find no reason to deny the kind of ethics and experience Taylor describes while at the same time moving to what might be seen as a reductionist historical explanation, one of the sort that Kitcher tries in his essay here, and that certainly is part of Wilson's imagination of religion and enchantment. Wilson and De Waal certainly, Costa and Richards also, though they go very different ways, seem to me to share this broad sense of possibility.

Like Taylor in his characterization of the secular position, though coming from a very different and quite Kantian direction, Richards argues that it is the human mind that "formulates the theories that articulate and characterize nature." And he reminds us that when "the human mind disappears," so do all of nature's "variegated colors, sounds, tactile feelings," and the wonder and splendor of my red-tailed hawk in Central Park. All of which, for Richards, leaves the world enchanted after all, just because "at a very deep level" the human mind serves as a "template for understanding nature," though the enchantment resides not strictly in the world outside mind; the source of enchantment is ultimately the human mind that makes sense of nature in the first place. To put it in other terms, those, for example, set by Paolo Costa in his reading of Heideggerian thought on these matters, it is a mistake to think that the issue can be understood entirely in the movement from inner to outer, mind to nature. Feelings themselves, "moods," Costa claims, are conditions for "having a world" at all.

I don't want here to engage readers or the other contributors in such complex philosophical arguments. They are important, and I'm hoping that readers will work through these essays and their arguments as they refuse simplifications, even perhaps the unwitting simplifications I must be making here. The detailed arguments and contentions, the tests of my ideas, and the interaction and conflict of others are what will make this book worth the trouble. But it seems to me essential to recognize that Taylor's argument about secularism's relation to a new subjectivity about the source of value is implicitly confirmed in many of these essays and is critical for our sense of the value of "this world," as I keep insisting on calling it. Did Darwin, for example, really love his mindless world? Well, I believe he did, but Richards shows that part of the reason for this may be that the world for Darwin wasn't really mindless after all; it is the contemporary mind that finally and fully evacuates mind from natural selection. But that mindless meaning is, ironically, entirely mind-made and thus may not be what the nostalgic yearner's return to the fullness of enchanted cultures requires; it does, however, suggest that the urgent requirement for a universal and objective set of strong imperatives and values is not necessarily the way to go as we wrestle with the question of value's place in a secular world. I would, for example, point to Taylor's formulation of the idea of "overlapping consensus," in his earlier essay on secularism. There is no way, even in a secular democracy, to make all the factions agree. The unity of a democratic culture, however, depends on an "overlapping consensus," places where each of the very different sects, factions, religions, social groups will agree, and thus places where the secular authority that binds them can find secure support.

This is, as I claim at the start, a secular book. Its aim is to move toward a rethinking and refeeling of secularism, repudiating the supernatural, and it works from a position that I was delighted to see formulated by Kitcher: "departures from naturalism are both unconvincing and antithetical," while at the same time—as certainly Taylor would agree—"naturalistic explanations of ethical practice are inadequate and fallacious." But much of what I have been saying here, even invoking Robbins's progress narrative and Wilson's distinction between pragmatic and true truth, implies a commitment to the moral importance of that "true" truth. As Kitcher, again, puts it, "the transmission of ethical precepts depends on the ability of the recipient to have confidence that the source of the precepts is good, and that already presupposes a capacity for independent—secular—judgment of the good." And that in turn implies some canon of reasoning and of evidential proof that will allow the identification of what's reliable. In other words, put simply, before appropriate action can be taken, one needs to know what's what—and here, I affirm once more, scientific canons of proof become essential.

This holds true at the higher levels of the question, whether, for example, humans can be imagined as "altruistic," and as ethical beings. Is morality an entirely human invention? We have seen that Frans de Waal, working with primates, helps me toward the answer I have been seeking. If it is true that we are creatures intrinsically incapable of real morality, the ethical questions I think we all need to raise become irrelevant. So, as with any good secular writing, it's necessary to get as close as we can to the fact of the matter. In this respect, ethics and science, the "ought" and the "is," are directly related to each other, at least in that ethics would not make any sense at all if nature makes us intrinsically incapable of moral principles and behavior. But all of us, even the most cynical, have evidence that we are not incapable, and thus the explanation must come either in reference to the supernatural, a more traditional response that even Alfred Russel Wallace, the codiscoverer of natural selection, came to accept in relation to the presence of human consciousness and spirit; or there is something in the nature of natural processes that, without transcendental assistance, makes the ethical possible. We know where Darwin stood on this. De Waal makes a firm case that study of primates provides unequivocal evidence that the ethical is built into the evolutionary process. It is certainly built into the nature of primates. Nature, it seems, is not only, not merely, red in tooth and claw even as—so Richard Dawkins enjoys reminding us—in the very instant in which you are reading this, millions of creatures are being pursued, devoured, destroyed, and brutalized.

That it is mind that, in a certain sense, creates the world (half perceived, half created), or at least creates all value, ought not to be an impediment to the development of ethical canons. Kitcher makes a strong case for considering the evolution of ethics because he sees ethical imperatives as a product of history. This surely will run into some trouble, in part because one might encounter conflicting versions of history (Kitcher's sketch of that history—but he sees it only as a sketch—sounds rather like the standard one developed by evolutionary thinkers in our own day, and one that could seem reductionist in just the way Taylor does not accept).

But he is surely right that "[e]thical practice . . . has a long history," and that one needs, as a secularist committed to the deep significance of ethical theory and practice, to use the perspective developed from historical study to "determine [ethical practice's] ability to yield objective claims." So for Kitcher, one begins by trying "to understand ethical practice, as one understands religious belief, as a historical phenomenon, to consider its evolution." Where Taylor turns away, Kitcher is ready, as he claims, in the light of possible evolutionary and adaptionist history of ethics, to see ethics as "a form of *social technology*." Again, not a charming or enchanting formulation, but one that it might be possible to work with. Where Taylor is understandably uneasy about *reducing* ethics to so pragmatic a matter, Kitcher, insisting on history, is ready to watch ethics grow, and I would say detach from its history in the sense that while it may in fact have developed in consonance with religion, there is no particular reason why it must be seen as indissolubly involved in it. To make a parallel argument: the development of Darwin's theory was certainly partly spurred by his reading of Malthus, and thus it was certainly comfortable for Darwin at least partly because, as Adrian Desmond and James Moore,[22] among others, point out, his theory was so much like the laissez-faire economics in which he participated cozily. Yet the theory surely does not now depend on the validity of Malthus's theory nor on the workings of capitalism. Evolutionary science has taken the theory a long way forward and recontextualized it. It remains enormously useful, but you don't have to be a capitalist to believe it. You don't have to be religious to accept ethical imperatives.

One won't, then, get to Taylor's universal and objective imperative (the imperative that inevitably, on his account, contradicts the mind-centered condition of secular value). But one might approach (asymptotically, I think, is the word, implying a "progressive" but never completed narrative) a working and utterly necessary ethics. And along the way, there would never be an invocation of the witch doctor's judgment.

So secularism isn't easy, and the danger of slipping into triumphalist rhetoric, instead of Woods's fraternal and sympathetic recognition of a religious alternative, is great. I would like to think that none of the contributors here indulge that rhetoric. All of them recognize, in the quest for a fully satisfying secularism, how deep in our moral and aesthetic traditions religion lies, but all of them are driven, as well, by a profound sense of the necessity of facing—and I use the word, if nervously, again— the truth. Take a full look at the worst, said Thomas Hardy, for that is a condition for getting it right and maneuvering this difficult, terrifying, and wonderful world humanely if not (certainly not always) successfully.

஑

It has been striking to me in reading the very various essays written for this volume that almost all of them arrive at points where the strong latent feeling surfaces—not that necessary impersonality toward which Weber wants science to aspire in his "Science as a Vocation," but an excited affirmation of things of this world, a feeling for them. And this feeling isn't for my all too symbolic but very real birds, of course (that's my eccentric problem), but for what Paolo Costa here describes as the world's startling, wonderful ordinariness, for the necessary generosity entailed in thinking about things shared and needed here and now. I don't want to force the contributors here to be saying my "yes," but I do want to call attention to the fact that these questions of "secularism" and "enchantment" are not *merely* philosophical, not *merely* theoretical, not *merely* historical.

The "care for being" that Connolly arrives at in his imagination of secular enchantment is one that, he says, "can be joined to political militance," but that will also "be affected by the sensibility infusing it." It's a "practical wisdom" Connolly seeks in his argument with Augustine as in his essay here. And it is a practical wisdom that I hope lies behind all these secular arguments for caring directly and primarily for the things of this world. "Life," Connolly points out, "need not be devalued because it has evolved from nonlife and is now irreducible to it," and so, too, "the human species need not be devalued because it has evolved from other species." Part of the point of this book is to insist that however we read the narrative of disenchantment, in this secular world there is no necessary devaluing of life only because in most cultures, ethics and art and value itself had been so integrally related to religion.

These historical, philosophical, and theoretical questions that play around the idea of secular enchantment are thus always also implicitly political. In requiring attention to conditions here and now, they give philosophy, theory, and history a practical and political and deeply ethical life—they are questions that matter to us now as we engage this world that (perhaps) we never made, or that we have made too fatally. They have as much political as religious significance, and that generic distinction within secularism must Darwinianly fade. (It is no accident that, if in very different ways, the essays of Kitcher, Robbins, Connolly, and Wilson become concerned, as they make their secular cases, with questions of justice and equity and freedom.) Is life worth living? Of course. These questions matter to us now and should be felt on the fingertips just as I feel the thrill and importance of that Central Park hawk or the stooping peregrine falcon, or the shame of the scandal of the homeless on the streets of New York, or of the misery and starvation and abuse of large portions of the population, signs of which I can see right in the Central Park Ramble that yields me so much pleasure and such a deep sense of fullness every spring migration, or of the disasters everywhere that somehow, luckily, haven't hit me. It's only luck, but that's okay too.

But is life worth living? You bet! The sheer joy of life, perhaps all too aesthetic and insufficiently other-directed,[23] that those birds offer me can manifest itself in many ways, but beyond such vital experience, perhaps growing from it, there are even more generous possibilities. In a personal note, Bruce Robbins summarized for me just the arguments I have been hoping to build to here: "Life is worth living because we have reason to think we have treated each other relatively well in the past, in spite of everything," and it is worth living "because we believe we can learn to treat each other better in the future, that doing so will not violate or contradict our inmost nature, that the move away from supernatural/enchantment will help us do this." The world counts first and we can take advantage of our luck to watch the stooping hawk and the flocking starlings as we learn more about it and make it better.

# I

## ❧

# Challenges for Secularism

*Philip Kitcher*

Secularism,[1] as I shall understand it, claims that there are no supernatural entities, nothing that fits the admittedly vague characterization of "the transcendent" to which William James reluctantly appealed in his effort to "circumscribe the topic" of religion.[2] More straightforwardly, secularists doubt the existence of the deities, divinities, spirits, ghosts, and ancestors, the sacredness of specific places, and the supernatural forces to which the world's various religions, past and present, make their varied appeals. For the past two centuries, a combination of scholars, working in many different disciplines, have articulated the challenge *of* secularism, a sustained argument, rarely presented as a whole, that makes belief in supernatural beings untenable. Although I shall start with a précis of this line of reasoning, my principal interest will lie in understanding the challenges *for* secularism. An adequate response to these challenges requires moving beyond secularism as a merely negative doctrine, and offering something to replace the functional aspects of traditional religions. Secularism needs to become secular *human*ism.

Despite the contemporary attention given to religious belief, both by militant assailants—"Darwinian atheists," as I shall call them[3]—and by defenders who want to claim the consistency of belief in the supernatural with everything we know, the state of the controversy strikes me as quite unsatisfactory. Darwinian atheists, among whom I include Richard Dawkins, Daniel Dennett, Sam Harris, and Christopher Hitchens, neither offer the best arguments against belief in the supernatural nor pay much attention to the challenges *for* secularism: it is enough for them to demolish, and they pay too little attention to questions that might arise for erstwhile believers after the demolition is done.[4] On the other hand, the would-be reconcilers do not face up to the most serious reasons for doubt about their favored transcendental being—typically, the Christian God—rebutting the oversimplifications of Darwinian atheism instead of addressing the challenge of secularism. Even the most subtle and many-sided attempt at reconciliation, offered in a long historical study by Charles Taylor, explains the displacement of religious belief by complex cultural processes that contrast with relatively simplistic accounts about the growth of rationality.[5] Taylor's discussion will be particularly important for me because of his sensitivity to the challenges *for* secularism, and his clear articulation of them. Before his concerns can be stated and taken up, however, there is significant ground clearing—Lockean underlaborer work—to be done.

## II

Many people have no difficulty in reconciling their religious commitments with the picture of the natural world disclosed by the natural sciences. Faced with evidence, available since the early nineteenth century, to the effect that the fossil record of life on earth is incompatible with the creation stories told in Genesis, they declare that these parts of the scriptures are myths, whose significance is moral and spiritual, not cosmological. Presented with a Darwinian account of the history of terrestrial life, they acknowledge the suffering and the wastage, but insist that it is not the place of finite creatures to assess the purposes of the Almighty. Confronted with "philosophical" objections to traditional "proofs" of the existence of God, they explain that their own beliefs never rested on sophisticated (possibly sophistical) forms of reasoning. More ambitiously, they may also gesture toward the writings of other—Christian—philosophers who chop the logic with even more skill than do the critics, and who allegedly show that the supposed objections are not decisive.

All this is beside the central point. It is a sideshow to the many-sided challenge *of* secularism, developed from the eighteenth century to the present (although there are earlier roots in thinkers like Hobbes and Spinoza).[6] The challenge starts from the question of what the basis for religious belief might be. One obvious possibility is that religious belief is acquired as people grow up within particular cultural milieus, typically absorbed from parents and teachers, occasionally (and only in relatively recent epochs) adopted more self-consciously by acquaintance with some other movement present in the social surroundings. Although this is not the only possibility to be considered, it is common enough to make an appropriate starting point.

The core challenge of secularism is an argument from symmetry. Variation in religious doctrine is enormous, and central themes in the world's religions are massively inconsistent with one another. Defenders of supernatural beings can sometimes conceal the difficulty from themselves by focusing on a few religions with shared central doctrines inherited from a common origin—as, for example, when religious diversity is conceived in terms of the differences among the three Abrahamic monotheisms. More radical problems emerge once one recognizes the possibilities of polytheism, of spirit worship, of the devotion to ancestors that pervades some African religions, of the sacred spaces of aboriginal Australians, of the *mana* introduced in Polynesian and Melanesian societies. Adherents of these rival views of the supernatural realm come to believe in just the same ways as do their Abrahamic counterparts. They, too, stand in a long tradition that reaches back into the distant past, originating, so they are told, in wonderful events and special revelations. Plainly, if the doctrines about the supernatural favored by the Christian—and also by Jews and Muslims—are correct, then these alternative societies are terribly deceived. Their members have been victims of an entirely false mythology, instead of the correct revelation lavished on the spiritual descendants of Abraham. What feature of the Christian's acceptance of Jesus as Lord and Savior distinguishes that commitment as privileged, marks it off from the (tragic) errors of the world's benighted peoples?

Nothing. Most Christians have adopted their doctrines much as the polytheists and the ancestor-worshippers have acquired theirs, through early teaching and socialization. Had the Christians been born among the aboriginal Australians, they would believe, in just the same ways, on just the same bases, and with just the same convictions, doctrines about the Dreamtime instead of about the Resurrection. The symmetry is complete. *None* of the processes of socialization, *none* of the chains of transmission of sacred lore across the generations, has any special justificatory force. Because of the widespread inconsistency in religious doctrine, it is clear that not all of these traditions can yield true beliefs about the supernatural. Given that they are all on a par, we should trust none of them.[7]

Two centuries of research in the textual analysis of scriptures (particularly the Hebrew Bible, and the Christian Old and New Testaments), in the historical understanding of the formation of religious canons, in the historical study of the political contexts in which religions have evolved, and in the sociological investigation of the growth and spread of religions, have deepened the symmetry argument. Scholars in these areas, many of them devout, have offered rigorous studies that combine to show how the processes through which religions evolve, decline, and grow are not at all conducive to the acquisition, refinement, and accumulation of truth. From the very beginning, Christianity has struggled to elaborate a single uncontested account of the events of the life of its nominal founder— the opening of Luke's Gospel tells us as much. The Gospels we have are often inconsistent with one another, and the incompatibilities are plainly the result of decisions that attempt to include various early Christian communities, devoted to alternative scriptural stories.[8] We also know of far more radical departures from the standard Gospels, accepted by other first-century movements that have been written out of the canon. Historical knowledge of some of the people who figure in Jesus' life— Pontius Pilate, for example—enables us to recognize as fiction some famous episodes: the invention of the idea of a Jewish mob that called out for Christ to be given a *Roman* punishment.[9] Sociologists have explored how religions appeal by offering to meet the psychological and social needs of converts—and how successful religions adapt by adjusting doctrines to meet these needs. There is no doubt that the traditions through which religious ideas come down to the contemporary world show an evolution of those ideas, and that the processes through which that cultural evolution goes forward are completely unreliable with respect to generating and spreading truth. Hence it is not simply that the ways through which religious people come to their commitments are equivalent across massive inconsistencies in claims about the supernatural, but the ways themselves, when scrutinized, turn out to be quite unconnected with the generation of true belief. Some religions succeed at propagating themselves across the centuries, but success has nothing to do with the determination of what any supernatural world is like—religions do not have to be correct to be widely believed.

A natural response to the challenge, as so far articulated, would be to maintain that the believer has been portrayed as far too passive, a mere recipient of the commitments that some surrounding culture impresses on its children. Some people, of course, change their religions, coming to accept as adults doctrines quite different from those they heard from their teachers and parents. Many, perhaps all, of these cases involve experiences that the subjects would describe as religious. Strange things are seen; voices are heard; there is a sense of some presence beyond that of the everyday world. In the most dramatic forms—Saul on the road to Damascus, for example—there is a sudden, violently produced, change of perspective. Other devout people would describe very different experiences, the constant awareness of their god (or God) as they engage in particular activities, prayer or ritual devotion, for example. Religious experience provides an independent route to religious conviction, one that sometimes generates something radically new, sometimes confirms the faith acquired early in life.

Unlike other modes of perception, including the special capacities of the hearer with absolute pitch or the accomplished wine-taster, religious experience cannot be assessed for its reliability through verification that it indeed does provide information about the entities to which it is supposed to give access. It is invoked precisely because there is no independent way of finding out about those entities. Some things about religious experience are, however, very evident. First, the statistics of reported religious experiences are highly variable across groups and across periods of time. Second, these experiences are more frequent when people have ingested substances that we do not normally regard as increasing their epistemic competence. Third, in their dramatic forms, they occur disproportionately among those who are anxious, fearful, disturbed, and distraught—again, prior states not usually viewed as ideal for observation. Most important, however, is the fact that any religious experience has to be *characterized*. The categories in which a religious experience is framed typically are those of the religion to which the subject is already committed: it is no accident that Christians feel the presence of Jesus, while ancestor-worshippers sense the proximity of their forebears; Catholics see the Virgin in a window in Brooklyn, but Protestants do not. Moreover, in all instances, the experience is classified in religious terms that are already familiar. Instead of being an independent test of the doctrines transmitted by tradition, religious experiences already *presuppose* the legitimacy of some traditional doctrines.

Even when the obvious worries about their genesis are ignored, religious experiences cannot break the symmetry among rival traditions. Reading the New Testament, the Christian claims to sense the divinity of Jesus. Equally, in the sacred places, the Australian feels the reverberations of the Dreamtime, and the Hindu at prayer has an awareness of some particular deity. The difficulties of escaping the hold of tradition are most evident on those occasions when a visionary claims to report something genuinely new. Religious traditions themselves have to grapple with the distinction between real insight into the supernatural realm and deceptive experiences that potentially corrupt the person and the religion itself. No wonder that many traditions have elaborate procedures for certifying those who would claim to have recurrent visions—as, for example, in the medieval processes for deciding whether the claims made by a would-be anchorite are genuine. How is the risk of possible heresy to be met? Only by insisting on conformity to the orthodox doctrines, as they have been passed down and recognized by those who are most qualified: the anchorite-to-be must submit to the judgment of priest and bishop, ultimately to the hallowed wisdom of the Church itself.

A second way of attempting to evade the challenge would be to abandon the attempt to defend some particular religious view of the supernatural, in favor of a weaker commitment. Perhaps the pervasiveness of religion among human societies can be seen as evidence for the existence of *some* supernatural entity (or entities). This approach would seek the shared content of all the world's religions, the lowest common denominator, as it were. Radical differences in accounts of the supernatural force agnosticism about specific doctrines: it is impossible to warrant the existence of the Christian God, or of Allah, or of the pantheon of Hindu deities, or of the Dreamtime, but, in their various ways, all religious people are on to something. Religion in general may be sustained, even though all the specific religions fail.

Reflection on the unreliability of the processes through which religions grow and flourish should already induce worries about this line of reasoning. From what we know of the evolution of religions, there is an obvious alternative to the hypothesis that the various religions have, in their different and differently inadequate ways, grasped a core insight. Successful religions meet psychological and social needs, responding to human anxieties and yearnings, binding people together (one popular etymology links "religion" to *religare*, to bind together). Members of societies without religion are likely to be converted to religious belief, and those societies are likely to be less cohesive and vulnerable to invasion. Religion would thus be prevalent because cultures that lack it or lose it tend to disappear, not in virtue of the fact that most people share a human ability to obtain a dim grasp of the supernatural realm.

The cultural evolutionary hypothesis just sketched needs far more careful elaboration and confrontation with historical and sociological evidence if it is to earn our assent. Nevertheless, it reminds us that the substantive doctrines of the various religions are extensive myths, made up to answer to psychological and social purposes. Once that is admitted, as it is indeed conceded by the aspiring champion of the-supernatural-in-general, we must ask what core insight can be retained. Secularism does not suppose that our current scientific understanding of the world should be certified as complete. The history of the natural sciences has been full of surprises, and later researchers have had to introduce types of things very different from those thitherto recognized (think of the development of ideas about matter, as atoms give way to electrons and protons, to a host of fundamental particles, to quarks, and possibly to strings, branes, and higher-dimensional spaces). Perhaps, at some period in the future, inquiry will disclose novel entities that our descendants can recognize as satisfying whatever marks were previously supposed to distinguish the transcendent. Perhaps. It is not an option that can be ruled out, although one may wonder how large a probability we should assign to it. Secularism is atheistic about the substantive claims concerning the supernatural offered by all the religions ever devised by human beings, but it should be agnostic about the claim that something legitimately characterized as "transcendent" or "supernatural" exists.

Given the argument from symmetry, it would be premature to label the potential supernatural entity as "god" or "spirit" or "force"—or, indeed, as "mind" or "creator" or "intelligence." Once religion is seen as the lowest common denominator of the various substantive faiths, there is nothing that can be deployed to describe the supernatural. Even if it exists, it is "something we know not what," and we delude ourselves by grasping at familiar concepts—those of our historically generated myths—to cover our ignorance.

In formulating this conclusion, I have introduced a word whose absence in previous discussions may appear to undermine the challenge of secularism. To think of religions as faiths might be to separate them from systems of belief that aspire to be counted as knowledge.[10] If religion is a matter of faith, then the question from which secularism starts can, it seems, be repudiated. Asking after the grounds of contemporary religious belief, and embarrassing the believer by demonstrating that the processes that underlie it are unreliable, tries to confine devout people in places where they do not belong. Lack of epistemically secure grounds can simply be conceded. No champion of any religion should be perturbed by the thought that he or she cannot provide marks that distinguish the preferred beliefs about the supernatural from those offered in rival traditions, or worried by the fact that religious traditions evolve in ways that have nothing to do with truth. To be sure, if religion were a form of knowledge, these considerations might be unnerving. That, however, is to mistake the character of religious acceptance. Properly understood, it is a matter of faith.

To retreat in this way can be very tempting for the beleaguered believer. Yet it is important to understand what has been given up. To begin with, believers in each tradition must recognize that the same ungrounded commitments are available to their rivals—and, while each may believe that the others are in radical error, there are no evidential grounds to support that verdict. Ungrounded acceptance of supernatural entities, indeed of a rich body of lore about such entities, is simply legitimate for anyone, and even the lucky ones with the correct faith will have no justification for thinking that they are right. How will their bold acts of faith affect their lives, and the lives of others? Perhaps they will be cautious, not permitting the particular doctrines to which they have committed themselves to have any impact on their decisions, intentions, and actions. If that is so, then the commitments themselves amount to no more than a motion of the lips, an affirmation that plays no role in the life of the "believer." If, however, more force is given to those commitments, if they are allowed to shape the decisions that are made, for the religious person and for others too, then the stance is morally dubious. Responsible action cannot proceed from beliefs that are adopted groundlessly, through wishful thinking, arbitrary choice, or a "leap of faith." Christians who turn to their Bible for guidance and support take their reading to improve their moral deliberations, precisely because they have been given God's Word, but, by hypothesis, that is an article of blind faith. The moral doctrines of their favored text may be excellent (or they may not), but the crucial point is that blind believers have no basis for attributing any authority to them. If others turn to different texts, actual books like the Qur'an or the Bhagavad Gita, they may do so with equal legitimacy. Even someone who makes a "leap of faith," and takes *Mein Kampf* or *The 120 Days of Sodom* to be divinely inspired, is equally warranted in basing his decisions and actions upon that text. When this point is appreciated, the danger of substantive commitments without evidence becomes obvious. For a religious believer to be morally justified in adopting doctrines on blind faith, the claims accepted must be so tightly circumscribed, so carefully sealed off from any role in practical affairs, that the commitments themselves dwindle into meaninglessness. Like the Cheshire Cat's smile, they linger even after the beast itself is gone.[11]

## III

From my brief précis of the challenge *of* secularism, I turn now to the challenges *for* secularism. Darwinian atheists think that once the case against the supernatural has been made, their work is done (and, of course, they think that the case is made rather differently from the way in which I have presented it). People should simply stop believing the myths about the supernatural with which human beings have consoled themselves, and in whose name they have done a wide variety of hideous things to one another.[12] Some Darwinian atheists even seem to believe that, if they ridicule the myths sufficiently loudly and sufficiently often, erstwhile religious people will be shamed into behaving like grown-ups, throwing away their faulty crutches and signing on to the great scientific adventure. Current evidence does not support that hope. In my English youth, nobody made much fuss about Darwinism or the theory of evolution—only a few eccentric extremists were prepared to suggest that the orthodoxy about the history of life was mistaken. According to the latest surveys, however, around half the British population now harbors doubts about Darwin. By the same token, God's stock seems to be rising. Yet these trends are occurring in the homeland of the most eloquent (and biologically well-informed) of all Darwinian atheists.

Religion will not go away simply because people are told—very firmly—that Proper Adults should have no truck with supernaturalist myths. Darwinian atheism accepts, and reinforces, a common assumption about religion, to wit, that being a religious person or living a religious life is primarily a matter of believing particular doctrines. Sophisticated thinkers about religion have, for a very long time now, taken a rather different view. Central to the religions of the world are many other things: complexes of psychological attitudes (aspirations, intentions, and emotions) among their adherents, forms of social organization, rituals, and forms of joint behavior. Within contemporary religions (and, for citizens of the affluent nations, most prominent in Buddhism, Judaism, and Christianity) there are movements that emancipate themselves from doctrine entirely: these forms of religion are simply not in the (literal) belief business. In their recitals of ancient texts, they recognize valuable stories, not to be understood as literally true but important because of their orientation of the psychological life, the pointing of desire in the right directions, the raising of some emotions and the calming of others. One might even conjecture that the social and affective aspects of religion were, somewhere in prehistory, the ur-phenomena of religion, that religious life begins with particular emotions (awe, joyful acceptance) and with shared forms of ritualized behavior, and that the stories Darwinian atheists wish to debunk are later supplements, devised to bind the earlier practices together.[13]

It is now possible to see how the assaults of Darwinian atheists can be ineffective, even counterproductive. They sing beautifully to the choir.

Those who have already learned how to integrate their aspirations, intentions, and actions with a disenchanted[14] view of the world will be pleased to think of themselves as behaving like Proper Adults, but most of those who have not taken this step will wonder how they are to live once the supposed "myths" they have adopted have been repudiated. Where will they find another form of association that brings them together with their fellows in ventures they can accept as important? How will the hopes that are currently bound up with their benighted supernaturalist doctrines be sustained once the connecting framework of belief is excised? Darwinian atheists overlook the fact that religions serve psychological and social functions, *even though on the explanation of the prevalence of religion offered by the challenge of section II this fact is central to the growth and spread of religions.*

To acknowledge the many-sided character of religious life, and to appreciate the ways that simply removing religious belief can disrupt people's lives, is not to fall into the trap of supposing that it would be better for "the masses" to retain their comforting illusions. Secularists should not patronize others by supposing that the bracing repudiation of the supernatural is only for the brave few—among whom they can, fortunately, count themselves. The important point is to appreciate the problem, to understand the ways in which, for many thoughtful devout people, the subtraction of literal belief about supernatural beings comes as a deep threat. Once that problem has been recognized, once it has been studied and its dimensions mapped, then secularism can evolve, as religions have evolved, to adapt to human needs. Simply condoning the old myths is by no means the only way to undertake that task. Another approach would be to inquire into the character of the perceived threat, into what exactly the believer thinks would be lost in abandoning belief, and to articulate secularism as a set of positive responses to those potential losses. To follow that path would be to transmute secularism (as blunt denial) into secular *humanism.*

The challenges *for* secularism arise initially from skepticism that anything can make good the losses it entails. It is insufficient to declare, as Darwinian atheists sometimes do, that the skepticism is groundless. "Look around," they exhort, "and you will see plenty of contented thoroughly secular people, living lives that satisfy them, joined in functioning and peaceful communities." (You will see this especially clearly, it is suggested, if you spend time in Scandinavia.)[15] *Correct as these observations are, they do not yet explain to those who feel threatened just how the reintegration of life-after-myth is supposed to go.* Perhaps examination of the lives of contented nonbelievers will disclose patterns from which an elaboration of secular humanism may emerge. Until the construction has been done, however, simply gesturing at people who are reported to have come to terms with mythlessness is unconvincing. For the religious person may wonder how that allegedly happy state is to be achieved, or, indeed, whether the reports of its felicity are not greatly exaggerated.

The issues to be addressed in constructing secular humanism are partly social and partly matters of individual psychology. Because I take the latter to be the deeper sources of concern, most of the remainder of this essay will be directed at them. In the rest of this section, I shall look briefly at some social functions of religion.

Statistical analyses indicate that religious adherence and religious fervor are strongest in those societies (and among those groups within societies) that are most vulnerable to the vicissitudes of life.[16] For people whose lives are going badly, or that are in constant danger of going badly, religion can provide important forms of security, sometimes hope that the reversals of this life will be compensated in the next, and opportunities for mutual consolation. Part of this promise (the idea that the bad things that actually occur will somehow be redeemed) is not easily replicable in a secular framework, and this is a point that will have to be addressed in the next section. Other aspects of religious reassurance to the endangered are, however, tractable through improvements in the social structures in which people are embedded. Some societies provide buffering against many of the major calamities of human existence, reducing the threat of sudden unemployment and penury, providing health care and support for the disabled, even taking steps to ensure that nobody is destitute and homeless. Humane social measures that take into account the needs of all citizens can easily substitute for the charity and material support provided by religious organizations.

Lack of assurance about one's continued ability to satisfy physical needs is only the simplest type of condition to which religions may be perceived as making a welcome social response. For people who are marginalized in society, who recognize that the existing secular institutions treat them unjustly, participation in religion may be bound up with a struggle to obtain the rights currently denied to them. Religious communities have often played an important role in bringing the powerless together, in combining their voices so that they can at last be heard, in providing a sustaining sense of solidarity that is expressed in courage and determination. Famously, the civil rights movement of the 1960s was grounded in the churches, and led by eloquent preachers who could galvanize their congregations. Less evident to many (although not to those who live in the northern parts of Manhattan) is the social role that churches continue to play in the lives of poor Americans belonging to ethnic groups that cope with the problems of long-permitted urban decay.[17] Religious communities continue to provide resources for families who struggle to create better opportunities for their children in environments where the secular institutions (schools, job-training programs, clinics) are frequently inadequate, and where the temptation to acquiesce in hopelessness is omnipresent.

*It does not have to be that way.* Secular society could in principle respond to the problems of social and economic injustice, so that the felt need for collective action or for a system of support, currently provided by the religious community, would already be met, as it is for more fortunate citizens. As a matter of fact, however, no response on any appropriate scale has as yet been given, and even when steps in the direction of greater socioeconomic justice are proposed, they are resisted by affluent legislators (who typically identify themselves as Christians). If secular humanism is to emerge from secularism, then one principal element in its positive position must be a firm commitment to increased socioeconomic justice, both within nation-states and across the entire human species, a commitment that is not simply a declaration of abstract rights but embodied in the sharing of the world's resources and opportunities in accordance with egalitarian ideals.

So far, the social achievements of religions have been considered merely as means, ways of providing for needy people things that secular institutions fail to supply. Thinking about religious communities in this relatively shallow instrumental fashion fails to identify their primary significance. Religious institutions connect their members, providing a sense of belonging and of being together with others, of sharing problems and of working cooperatively to find solutions. For many people, religious involvement provides occasions not merely for talk about important issues—although that itself is valuable to them—but also for conjoint action. Religious communities can come to agree on goals, not necessarily centered on the liberation or socioeconomic progress of the members, and can advance broader projects that are to be collectively pursued.

Again, *religion does not have to be the main vehicle of community life.* Thoroughly secular societies can have community structures that enable people to enter into sympathetic relations with one another, to achieve solidarity with their fellows, to exchange views about topics that matter most to them, to raise questions about what should be done, and to work together toward goals that have been collectively determined. Very probably, many of the authors of secularist manifestos are embedded in community structures of this sort—and perhaps the apparent naturalness of the social relations they enjoy hides from them the fact that similar *secular* structures are not available to others. For many Americans, however, there are no serious opportunities—outside of churches, mosques, and synagogues—for fellowship with all the dimensions just discerned. Perhaps, in groups of friends, there are occasional opportunities for more serious discussion, for self-revelation and confession of doubts, for exploration of what is valuable and what is to be done. Or, perhaps, in secular settings, that would simply be embarrassing, and so the necessary words are unspoken; the spread of sympathy into the lives of others is limited; goals are decided and pursued largely alone. A secular world would thus appear to lack the most significant parts of community life.

Secular humanism needs a diagnosis of the human need for community that is far more sophisticated than that sketched in the preceding paragraph. For an adequate rebuilding of satisfying community structure in a secular society will depend on a charting of all the various functions that religious communities, at their best, currently discharge. It has been plain for decades—possibly for centuries—that maintaining a sense of community in a large, diverse society, especially one dedicated to economic and social competition, is very hard.[18] It is a significant fact that some of the most successful American ways of building secular community structures open to all—Unitarian churches, the Society for Ethical Culture, Jewish Community Centers—have imitated features of religious life and practice.[19]

# IV

A familiar feature of those groups of people who are often charged with practicing pseudoscience is the zeal with which they wrap themselves in the social trappings of the prestigious natural sciences—forming "Institutes," holding "conferences," setting up "peer-reviewed journals," and the like. Critics contend that these activities are mere gestures, that the ardent imitators may talk the talk but that they avoid what is really central to scientific practice. By the same token, those who feel secularism as a threat are likely to wonder whether the social surrogates just discussed really address the losses that the repudiation of religious doctrine would entail. Like the constructions of the pseudoscientists, the secular social substitutes are hollow, offering the *form* of community and of conjoint ethical action, but lacking the substance. Reactions of this sort rest on the thought that, while material benefits may be offered, deep psychological needs are not to be satisfied in these godless ways. Without the acceptance of "transcendent" entities, people will never achieve genuine community with others, never penetrate to the really important issues that concern them, never have the possibility of combined ethical action. The worries that underlie this reaction are sufficiently varied that any essay-sized response cannot do more than indicate ways of addressing the major points of anxiety. In this section, I shall try, relatively schematically, to take up three of these. The final section will be devoted to a further issue, extensively discussed by Charles Taylor, whose focusing of the point will enable me to be more thorough in canvassing the success of secular surrogates.[20]

Begin with the thought that there is something special, and irreplaceable, about the forms of community produced by shared religious doctrine. Without the common acceptance of some transcendental entity, in whose worship we are joined, the bonds of fellowship are, allegedly, weakened. Why exactly should that be so? Can't people find connection with one another through their mutual sympathies, or through the sharing of a cause or an ideal? What exactly does the invocation of some supernatural being add?

One answer to these questions, natural for those who accept one of the Abrahamic religions, is that belief in a divine creator introduces a special form of fraternity. We are joined in brotherhood—or, better, linked as siblings—because we are all children of the same God. Although it is not immediately evident how this conception of shared parentage is manifested in the religious communities actually formed—since these groups are typically far narrower than the entire species, or even those members of it that acknowledge the same deity—that is not the principal difficulty with the proposal. Rather, a secularist can reasonably respond, the closest relationships people, whether religious or secular, enjoy with one another are grounded in complex forms of mutual sympathy. Why isn't the sympathy that leads one person to feel another's projects as her own, to modify her own plans and behavior to accommodate what the other person wants, to share in joys and sorrows, a sufficient basis for community, independent of any special type of common belief? Perhaps people who can form communities do need to reach agreement on some things, even on the most fundamental things, but there is no clear justification for supposing that that agreement has to take the very particular form of acknowledging a supernatural being.

Reply: the thought of a common relation to a transcendent entity, as in the idea that we are both (or all) the Children of God, goes beyond the mere accidents of human sympathy to endorse the special value that each individual has; in a religious community, you don't just see your fellow members as people with whom you happen to have sympathetic connections, dependent on the vagaries of your particular beliefs and wishes, but as loci of *real value*. So, for the devout, fellows in religion are not simply joined in mutual sympathy, but see one another as *worthy of that sympathy*. This response does add something that appears to distinguish the special quality of religious community from the secular approach offered in the previous paragraph, but its success depends on whether secularists can find a counterpart for the extra ingredient. The crucial question, then, is whether people who deny all supernatural beings are able to judge that some of their fellows are worthy of their sympathy and support.

If the considerations just adduced are correct, then suspicion about the possibilities for secular community rests on the supposition that certain kinds of value judgments are not available to nonbelievers. It is, of course, a common prejudice that repudiation of religion deprives one of any grounds for determining that anything is right or wrong, good or bad, valuable or worthless. Since I see no plausible thesis that would concede to secularists the possibility of making and defending judgments of value but deny the capacity for the assessments that figure in genuine community, I suppose that any denial that unbelievers can find their fellows worthy of sympathy and support would rest on the general complaint that, without deities (and the like), ethical claims become meaningless or unjustifiable.

Secular humanism should address that complaint in two ways. First, it should recall the point, familiar since Plato, that attempts at a religious foundation for ethics fail completely. Second, it should show how ethical practice, and judgments that certain things are worthwhile, can be understood in a thoroughly naturalistic fashion.

Religion may be thought to undergird ethics in either of two forms. First, as Euthyphro incautiously told Socrates, what is good may be *defined* as what the deities have willed.[21] Less ambitiously, one may propose not that goodness is stamped on some things by acts of divine will, but that a supernatural being *reveals* to us the independent character of what is good, thus supplying knowledge we would be unable to obtain in any other way.

Neither of these suggestions is defensible. The first encounters the dilemma Plato formulated. If the divine will is grounded in some apprehension of an independent standard of goodness, then that will is not the source of goodness. If, however, the divine will is not grounded in any such apprehension, the proclamation that certain things are good and others not is arbitrary fiat, and there is no basis for appraising the quality of the choices the deity makes.[22] In praising the deity, religious believers think of themselves as saying something substantial in extolling the goodness of God, not as simply asserting the consistency of the divine will.

The problem can be deepened through a consideration of the weaker suggestion that the divine role in ethics is one of transmitting to human beings the independent standards of goodness, standards that God can recognize but that are beyond our powers of apprehension. This approach opts for the limb of the dilemma that recognizes an independent source of value, but still finds a privileged role for the favored supernatural entity. Assume, then, that there has been some divine revelation, and that a particular group of people has been favored with God's commandments. If this transmission is to play a role in their ethical life, then they must view it as supplying reasons for doing the things commanded. They feel the obligation to accord with the commandments because a particular being has issued the commands. After the events of the twentieth century, and our familiarity with people who attempted to excuse the hideous things they did by declaring that they were following orders, we should be wary of the thought that, *all by itself,* an order can supply appropriate grounds for action. Is it appropriate for someone to follow any command from an outside source about what is to be done? The obvious answer, of course, is that it is sometimes proper to defer to others when one supposes that they are likely to command the *right* things. In the case of the deity, the commander can perhaps be recognized as having special attributes: this being is very powerful, knowledgeable about things to which lesser beings are blind, and so forth. Despite these impressive qualities, something crucial is lacking. Power and knowledge alone do not provide the kind of authority at stake here. Many of the underlings who were questioned about the atrocities they performed could point to the power and knowledge of their superiors. Their misdeeds stemmed from their willingness to follow commands from sources they could not justifiably think of as good (and, in most cases, should have judged to be profoundly bad). Deference thus depends on a capacity to recognize the source of the commands as good, and that requires just that ability to assess the deity, independent of his pronouncements, that the religious account denies. In short, the transmission of ethical precepts depends on the ability of the recipient to have confidence that the source of the precepts is good, and that already presupposes a capacity for independent—secular—judgment of the good.

It is now possible to see more clearly what goes wrong on the other limb of the dilemma, that is, if one supposes that the divine will *makes* some things good and others bad. Imagine that you are the recipient of the commandments, and that you know this fact about their production. You might follow the decrees, perhaps out of fear of the power of the divine commander, but your following them could not amount to any form of *ethical* life. Your attitude would be one of constraint, like that of a subordinate more thoroughly controlled than those who served the twentieth-century regimes of terror. The pursuit of *goodness* would have nothing to do with it, for any such concept would be inapplicable to your immensely powerful boss, and, in consequence, inapplicable to the edicts that he has issued.

Ethics is commonly embedded in religion because many people do not see how there could be any secular basis for judgments of value. Despite the fact that, under scrutiny, religion cannot provide any basis for ethical practice, the prejudice that it can endures, in part because of the implausibility of available philosophical accounts. Secularism is likely to remain suspect so long as no convincing explanation of ethical practice has been given.

Hence secular humanism should contain an extended answer to questions about our practices of making judgments of value, so as to show what these judgments amount to and how they are possible. In repudiating supernatural beings, it should be equally skeptical about the nebulous entities and processes—far stranger to ordinary folk than the traditional divinities—invoked by the philosophers: nonnatural properties, faculties of moral perception or ethical intuition, commands of pure practical reason and the like. Yet efforts to locate sources of value in the natural world typically founder, committing familiar fallacies or making crude identifications of what is good with what promotes evolutionary advantage.[23] It seems, then, that secularists face a dilemma of their own: departures from naturalism are both unconvincing and antithetical to the line of reasoning that repudiates the supernatural, while naturalistic explanations of ethical practice are inadequate and fallacious.

The way out is to emulate the strategy that figured in the argument against supernaturalism, to understand ethical practice, as one understands religious belief, as a historical phenomenon, to consider its evolution, and to use the resulting perspective to determine its ability to yield objective claims. Here, I can only indicate briefly how this approach might be pursued.[24] Ethical practice, I propose, has a long history, possibly a history as long as that of our species. It began when our ancestors, living in small mixed groups of the sorts found today among chimpanzees and bonobos, became able to go beyond the tense and fragile social lives of such bands by acquiring a capacity for *normative guidance*. That capacity consisted in a new ability to formulate rules for their own conduct, and to inhibit forms of behavior that would lead to social disruption (as well as trouble for themselves). At an early stage of *human* life, probably at least fifty thousand years ago, normative guidance became socially embedded, through the discussion and formulation of rules among the mature members of the social unit. Out of that primitive rule-setting practice came a variety of "experiments of living" (in Mill's famous phrase) that have been in cultural competition with one another. The ethical practices of today are the remote descendants of earlier efforts that were successful in this competition.

At the dawn of human history, with the invention of writing, we can begin to observe the precepts according to which societies were organized, and the fragmentary nature of the early legal codes makes it apparent that they are the heirs to systems of rules that had been developed much earlier—rules that had originally made the expansion of human social groups possible. For the past five thousand years, it is possible to recognize, still only incompletely, the further evolution of ethical practices. During this period, we can identify, if only occasionally, episodes of what seem like ethical advance: slavery is repudiated, women are given greater opportunities, people are no longer condemned for having sexual tastes that differ from those of the majority. By examining the processes through which transitions that appear progressive are effected, it is possible to investigate issues of ethical truth and ethical knowledge. If ethics is about anything, then clues to its content ought to be found at those moments where ethical advances are made.

On the account I propose, ethics is a form of *social technology*, originally introduced to address the fragility and instability of chimpanzee-bonobo-hominid social life. In its initial form, ethical practice was surely crude and simple, but forty-five thousand years of experimentation have accumulated solutions to the original difficulties, new problems posed by those first solutions, secondary solutions to those unanticipated problems followed by yet further challenges, and so on and on and on. Progress is to be understood in terms of the *evolution from* the initial state, not as *steps toward* some final ethical system. In this way, I believe, one can make sense of the human practices of valuation, of the traditions in which they are embedded, and of the hold they exert upon us. Like earlier pragmatists, particularly Dewey, I suppose that ethical practice is grounded in very basic human desires and in features of the human condition (specifically the forms of social life to which our prehuman ancestors were already adapted). Ethics is something *we make*, but we do not make it arbitrarily, for the conditions under which our ancestors made it and under which we continue to make it are determined by the species of animal to which we belong.[25]

If a naturalistic approach akin to the one just sketched can be given,[26] then the double-sided case against the necessity of religious foundations for ethics is completed. Secular humanism can thus draw on the negative arguments offered by Plato and his successors, without invoking a surrogate that is equally nebulous and far less appealing than the traditional connection of the ethical life to the dicta of the deity. It can thus turn back the challenge that genuine community and conjoint ethical action are quite impossible, once religion has been abandoned. Moreover, it can address a second issue often raised by religious people: the suspicion that lives without God (or gods) are deprived of purpose and significance.

There is an obvious sense in which secularism excludes purpose. If you suppose that the universe has been created according to a plan, then you can think of its history as unfolding that plan, as realizing the goals that the creator had in mind from the beginning (if, that is, such temporal language even makes sense in application to a supernatural being). Contemporary cosmology, and, even more evidently, the Darwinian account of the history of life, do not indicate any such purposive development—as we saw earlier (at the beginning of section II), recognizing life as evolving for close to four billion years, with natural selection as the main agent of evolutionary change, exacerbates traditional worries about the contrast between the divine perfections and the messiness of the world we inhabit. To say, however, that the secularist perspective eliminates purpose *tout court* would be a mistake, for, as with ethics, we can think of attributions of purpose as something human beings do. Purpose is not imposed on our world from the outside, by divine fiat, but *purposes* are made up by us. Purpose making is part of the human practice of valuing, a practice that has been with us for most of our history as a species. .

The sweeping claim that secularism removes purpose from the universe is mistaken, but the claim seems unnerving because of a very specific corollary. If there is no purpose (period), there can be no purpose to a human life—our existences become pointless. If that corollary is accepted, then religious people wonder, quite reasonably, whether the forms of community envisaged by secularists can ever touch the "really deep issues." Central to the questions they label in this way is the query, posed by almost all of us at some especially reflective moment, of what our lives are about, what gives them some sort of "meaning." However they fashion their communities, secularists (so the story goes) can find nothing to say in response to the query—or in response to it in its "deepest" form—because any adequate treatment of it would require there to be purpose, and purpose has been eliminated from the start. Secularist conversations, however hard they may strive to cope with the most intimate concerns, are forever precluded from supplying what the services and discussions among the faithful regularly sustain.

Again, secular humanism should address this challenge in both negative and positive ways. Can the external imposition of purpose on human lives actually achieve what the religious believer usually attributes to it? Is the human activity of making purposes for ourselves, of giving point and direction to our lives, somehow inadequate, missing some important feature that the religious versions of the significance of human existence supply? I shall try to show, as before, that the *human* practice of valuation is actually critical to our finding genuine meaning in our lives.

As already noted, secularism removes purpose from the cosmos, by denying that there is an unfolding plan, envisaged by a divine creator. The removal renders impossible a particular way of conceiving the significance of human lives. You can no longer think of your life as directed toward filling some small (probably infinitesimally tiny) part of the divine scheme of things. What is lost here is the thought that a specific task— perhaps that of recognizing the greatness of God and working to bring about his will on Earth, as is best in your particular situation—has been assigned to you, and that this assignment provides your life with direction and meaning. A first secularist response would deny that this is any loss at all, and would assert that, on the contrary, we *gain* significance for ourselves once we recognize the importance of choosing our own pattern and our own projects. The response opposes to the religious conception a vision of human individuality that sees the imposition of a particular role upon us as a denial of what is most important, namely, our capacity for choice for ourselves. In this juxtaposition of perspectives, two very different ideals of being human clash with one another: secularist emphasis on autonomy and self-choice contends with a religious ideal of humility and self-abnegation, submission to something far larger than oneself. Sincerely religious people may even decry the emphasis on autonomy as a form of arrogance, and as expressing the corruption of values in the transition to the Enlightenment (and post-Enlightenment) perspective.

It might appear that there is an inevitable standoff here, with each side committed to different views of what it means to be human, so that the debate about purpose and significance can never be resolved. On closer examination, however, the religious conception of how human lives obtain purpose tacitly presupposes that form of valuation on which secularists insist. For the considerations that undercut the idea of religious foundations for ethics arise here in a new guise. Believers do not think that the act of self-abnegation and submission to the will of the deity is *compelled*—and if they did, they would probably not view it as giving to a human life the significance they try to explain—but rather as issuing from the choice of the person who submits. A human life takes on its meaning because someone consciously identifies with the divine purposes, even when those are opaque. That identification must itself spring from an act of evaluation, a recognition not only of the power of the deity but of the *value* of serving a broader purpose accepted as *good*. Without such acts of evaluation, religious submission to the deity is far too close to the acceptance of those who commit themselves to serving the powerful agencies of the everyday world, who acquiesce in the commands of dictators, even when those dictators order the performance of atrocities. Religious people suppose that they can distinguish service to their deity (or deities) from the disturbing modes of blind self-abasement, and that power to distinguish already concedes the possibility of basing one's life on an act of evaluation. The difference between the two ways in which purpose is found does not therefore lie in the fact that the believer gives total priority to submission and humility over autonomous choice, but rather that *in an autonomous choice* the believer resolves to abdicate autonomy in order to serve what the autonomous assessment has already recognized as good. At bottom, both parties must accept the thesis that identifying purpose in one's life requires an initial decision to value one particular course, an act of validation made by an individual. The difference is that, on the religious account, the appropriate valuation is to embrace a purpose set down externally, one larger than that of any finite being, so that, at the first step, further autonomous decision is given up, whereas, according to the secularist, the purposes are thoroughly constituted by acts of human valuation. (There are, after all, no larger purposes to be embraced.)

Hence the religious challenge cannot be directed at the misguidedly presumptuous activity of human attempts to decide what is valuable, but must suppose that there is some quality of the purposes embraced by the believer that sets those apart from the projects that people can conceive for themselves. The language of the preceding paragraph already suggests what that quality is likely to be: divine purposes are cosmic, far larger than those human beings can set for themselves. Identification with God's will can be viewed as important *sub specie aeternitatis*. Not even the most ambitious secular projects that give shape to our lives can aspire to anything similar.

The question of what makes for a worthwhile life is as old as philosophy, and is, arguably, the central question of philosophy. Yet for a long period in Western thought, the period in which Christianity dominated, it hardly figures at all. Only in the late eighteenth century, as religion begins to confront the challenge *of* secularism, does the issue come, once again, to the fore. In ancient thought, accounts of the good life are often uncontaminated by any conception that a necessary condition for significant existence is some eternal contribution to the cosmos.[27] In the wake of Christianity, however, it is easy to be haunted by the thought that nothing less than a permanent imprint on the universe is enough. That thought combines easily with a view already prominent in the ancient treatments, the supposition that the good life is only for the few. Hence it is easy to conclude that worthwhile secular lives, even if possible at all, must be *exceptional*. Secularism thus falls radically short of the Christian promise, which effectively answers the question of the valuable life by declaring that it consists in service to God and the promotion of the divine purposes, and consequently takes significant lives to be available to all, even to people who suffer terribly.

Secular humanism should reject both the demand that genuine human purposes must connect to cosmic purpose and the exceptionalism that pervades ancient treatments of the good life. Individuals give their lives purpose and meaning by defining for themselves what matters most, shaping those lives around projects and relationships. If worthwhile lives are hard to achieve, that is not because people lack the opportunity to shake the universe, but because they are constrained in their choices— sometimes by the prejudice that what they must aspire to do is to "shake the universe"—or cut off from the opportunities that would be most ful-filling for them. Nevertheless, there are numerous lives, secular as well as religious, that find meaning in service to other people, in sustaining a family or a community, in working for the relief of the sick and the suf-fering, or in providing things that bring to others security or joy. Those who touch other lives most deeply may wonder if they have done enough, but their sensitivity to what further things might be achieved only testifies to the great value of what they have already done. To wish that one had done more is tacitly to endorse the value of the efforts seen as partial.

The dedicated nurse, the devoted family member who nurtures rela-tives, the indefatigable sustainer of some necessary public good—all these people, and many more, contribute to lives that extend beyond their own. Others reach further, making discoveries or creating works of music, art, and literature that bring inspiration, awe, and delight to many succeeding generations. Yet all these effects are thoroughly finite. There will come a time in the history of the cosmos at which all those immediately affected are dead, when the contributions are long forgotten, when the human species itself is extinct. So, a religious believer may maintain, these secular purposes fall short; in contrast to the larger purposes of the divine plan, they prove ephemeral.

That the effects of what we do are transient should be acknowledged. Is it, however, something that should be mourned? Is it a defect so decisive that it undermines any attribution of purpose to human lives? Secular hu-manism needs an articulated perspective that supports negative answers. Although I cannot fully present any such perspective here, I think it is possible to envisage the lines along which it would be developed. When our lives touch others in ways that protect them, or open up possibilities for them, we establish a connection with something that survives our own individual finitude. As when a stone is dropped into a pool, ripples continue after the stone has vanished, before they eventually die away. So, too, with our projects and our strivings, when they are well directed and well pursued. Even though the difference we make is not permanent, our having been affects something larger than our own existences, and thereby links us to a world that endures beyond us. We can abandon the hankering for the eternal contribution and still recognize the signifi-cance of the finite limited impact that we, finite limited beings, manage to achieve.[28]

Two main challenges have been considered: the thought that secular-ism can find no place for value, and the charge that it cannot identify purpose or significance in human lives. For my third challenge, I turn, far more cursorily, to a feature of the secular perspective that is easily connected with issues of purpose in human lives, particularly with human finitude. Religions sometimes, although by no means invariably, offer the prospect of immortality, and that offer may be understood as delivering forms of hope that secularists have to relinquish.

Is death fearful? Some thinkers have thought not, or at least not for those who have come to maturity.[29] Resentment or regret may be a more appropriate emotion, or resistance an apter attitude toward death, yet there may still be things about death that it is reasonable, even for fully grown-up people, to fear. We may be appalled by the gradual loss of abilities that is likely to precede the end, even the loss of the full vigor of our youth. We may fear the possibility that what remains in some terminal stage may not be anything we can identify as the self we value. We may be frightened of the pain of the process of dying. The most basic fear (or regret or resentment) is, I suggest, the anxiety that death will damage the value we aim to create with our life. What is most frightening is the prospect of *premature* death.

The religious offer of immortality helps with none of this. Whether we reemerge in some wondrous state after we die does not halt the dwindling of our capacities or preserve our youth, does not relieve us from that final stage of pitiable half-life (or less), does not salve whatever agonies come at the end. Nor can it complete the projects we leave unfinished. If your life is directed toward nurturing others who need your protection and guidance, and if, unluckily, you die before they are ready to cope without you, the fact that you will be restored—and maybe restored to them in some entirely different state—is immaterial. Your project, on which you have centered your existence, has still been compromised by premature death. Conversely, the prospect of death ceases to appear terrifying to those who have lived long enough to recognize that the central aims and purposes to which they have directed themselves are firmly in prospect, even fully secure. Their recognition does not halt the desire to live on, to see the course you have tried to direct unfold further, but, knowing that there is to be a terminus, you can become reconciled to its coming at any future point. The nurturing parent (and grandparent) sees the children (and the grandchildren) well launched in life, having found their own way and following it confidently, and would like to see further continuations of the family story—yet it is clear that, whenever death comes, there will always be loose ends, more episodes not yet seen.

The example I have chosen reflects the most common way in which people, religious or secular, find purpose in their lives, and it also indicates the most obvious context in which the religious promise of life after death brings hope. For those whose lives are centered on relations to others, death entails losses for which there can be no secular compensation. Although the deaths of young children are particularly poignant, the subtraction from your life of someone whom you love is painful at any stage—witness the decline of so many people who lose a lifelong companion. Here, apparently, religions like Christianity can offer hope—as Charles Kingsley wrote to his friend Thomas Henry Huxley on the death of the latter's beloved young son, he regretted that the famous agnostic (who had even coined the term) could not have the consolation of anticipating a reunion with the boy in the hereafter.[30] (Despite his grief, Huxley responded with an unflinching declaration of his resolve to "serve Truth.")

From conversations with religious people who do look forward to some sort of encounter with those whom they have loved and lost, I know how powerful this promise of hope can seem. Yet the sense of consolation depends, I believe, on not thinking hard about the terms of the offer. It is easy to suppose that the reunion will amount to a continuation of what has been, as if the tape of life were replayed and the death avoided. Plainly, that cannot be what occurs. Huxley could not have what he most wanted, the continuation of his young son's life: he could not see Noel grow up and find his own pattern of earthly life. Moreover, any reunion would apparently confront two strangers with one another, a parent whose life had extended in different directions after Noel's death and a child who would no longer occupy the emotional space vacated by his early death. Perhaps human understanding of the conditions of the reunion is defective, and the characterization just given is inadequate, but that is surely no help in providing consolation for the bereaved. It is common for Christians to disdain the material comforts of the paradise offered to Muslims in the Qur'an, but the Islamic vision of the hereafter does have the advantage of connecting with the desires of the faithful (the emphasis on flowing water is quite comprehensible, given the desert conditions in which the early community lived). Christian doctrines of resurrection, by contrast, are admitted to be mysteries, and, when that fact is appreciated, it is totally uncertain whether the conditions of our reunion in the hereafter make any response to the pains and losses we actually feel in the mundane world, or whether the form of existence envisaged is anything that would compensate for actual human grief and suffering.[31]

Many—perhaps most—human lives do not go well. Death often removes those we love, and shatters our projects. Secular humanism is committed to the attempt to decrease the frequency with which people's aspirations are frustrated and broken, despite recognizing that it can never expect to turn back all the reversals our mundane existence brings. Rather than promise some nebulous hope for the future, its attention is clearly focused on enlarging the prospects that the purposes we set for ourselves will be achieved, and on providing whatever consolation can be given when those purposes fail. The hope that is apparently abandoned is less wonderful than the religious take it to be, and, while life in a completely secular world is always vulnerable, it is not, on that account, bleak and hopeless.

# V

In his recent analysis of how we arrived at contemporary secularism, Charles Taylor plainly diverges from some of the conclusions of previous sections. His historical narrative views the potential basis for repudiation of the supernatural quite differently: instead of the challenge *of* secularism reconstructed in section II, Taylor considers only less powerful forms of reasoning, lines of argument that would not make a conclusive case against supernaturalist beliefs. Having substituted weaker reasoning, he is free to interpret the transition to secularism as occurring because of various social movements that make up for the deficiencies of the anti-supernaturalist arguments. Hence, Taylor can preserve a space for the—Christian—religious doctrines he regards as a live option.

Much as I admire the historical narrative, it suffers from underestimating the challenge *of* secularism. In presenting the challenge *for* secularism, however, Taylor avoids the popular versions of the charges, offering something more sophisticated. He would probably not accept the answers of sections III and IV, but it is clear that he does not take the crucial shortcoming of secularism to lie in the lines of criticism they address. He acknowledges forthrightly that secular lives can have a "moral/spiritual shape."[32] I take him to recognize very clearly that secularists can still find a place for talk about values and the purposes people attribute to their lives. Nor does his central concern about the secular forms of existence center on some loss of hope, enjoyed by religious people who can look forward to an afterlife. Rather, he insists, throughout his long book, on the idea that the secular life is "flattened," that it has, in a useful metaphor, lost a dimension.

As I interpret him, Taylor is concerned to elaborate a view toward which William James was groping in his classic discussion of Mysticism. James characterizes many mystical writings as "musical compositions," and he takes this "music" to supply "ontological messages which non-musical criticism is unable to contradict, though it may laugh at our foolishness in minding them." So we receive "whispers" from a world beyond, and, if we heed them, we come to "live in the eternal."[33] James struggles with the epistemological question of whether these experiences supply any warrant for belief in the supernatural, but he is thoroughly convinced that the conception of a larger realm beyond that of everyday (and scientific) experience enriches certain parts of human life.

Early in his book Taylor introduces the notion of a religious dimension in human experience, with a quotation from the Catholic writer (and monk) Bede Griffiths, a passage that serves to anchor later discussions.

> One day during my last term at school I walked out alone in the evening and heard the birds singing in that full chorus of song, which can only be heard at that time of the year at dawn or at sunset. I remember now the shock of surprise with which the sound broke on my ears. It seemed to me that I had never heard the birds singing before and I wondered whether they sang like this all year round and I had never noticed it. As I walked I came upon some hawthorn trees in full bloom and again I thought that I had never seen such a sight or experienced such sweetness before. If I had been brought suddenly among the trees of the Garden of Paradise and heard a choir of angels singing I could not have been more surprised. I came then to where the sun was setting over the playing fields. A lark rose suddenly from the ground beside the tree where I was standing and poured out its song above my head, and then sank still singing to rest. Everything then grew still as the sunset faded and the veil of dusk began to cover the earth. I remember now the feeling of awe which came over me. I felt inclined to kneel on the ground, as though I had been standing in the presence of an angel; and I hardly dared to look on the face of the sky, because it seemed as though it was but a veil before the face of God.[34]

The power of this example—"Bede Griffiths's Epiphany," as we might call it—lies in its resonances with the experiences of many people: the stage of life (around eighteen), the time of year (spring), the time of day (early morning or evening), the setting (familiar country under unusually beautiful light). For Taylor, however, the experience stands for something distinctive about the religious sensibility (this "one example" is to "stand for many"). It indicates that

> [s]omewhere, in some activity or condition, lies a fullness, a richness; that is, in that place (activity or condition), life is fuller, richer, deeper, more worth while, more admirable, more what it should be.[35]

What secular lives lack is not the values and purposes, the "moral/spiritual shape," but, Taylor suggests, this sense of fullness.

Powerful though the passage is, we need to be quite careful in distinguishing various questions that arise about it. Here are those I take to be most important:

1. Can thoroughly secular people have experiences like Bede Griffiths's Epiphany?
2. Do such experiences provide any evidence for a "supernatural realm"?
3. Are the experiences of those who believe in supernatural sources richer and deeper than the experiences of those who do not?
4. For religious people, do such experiences reinforce a sense of purpose and meaning in their lives, and do they strengthen the commitment to shared ethical projects?
5. Are experiences of this sort more readily available, and more sustainable, if they are linked to systems of religious belief?

Beyond the historical analyses, Taylor's book obtains much of its genuine power from suggesting answers to some of these questions, and letting them spread—by what I regard as unhealthy contagion—to others.

Begin with the easier questions: do secularists have similar experiences? Of course. Perhaps they happen above Tintern Abbey or in prospect of Mont Blanc or when the light strikes a particular Manhattan facade. Sometimes they occur in reading, or in listening to music. For thoroughly secular people, too, there come occasions of uplift, feelings of connection to others or to places, a sense that this is how life should be. There is no hint that these experiences are somehow "flatter" or bereft of some quality that religious people find in them. Possibly we secularists deceive ourselves, for even though we may try to compare our current epiphanies with those we enjoyed when we were once believers, memory may prove deceptive here. Let us, then, rest with the recognition that the experiences occur, and postpone the harder issue of a difference in richness or depth.

Question 2 already received its answer in section II, but it is worth revisiting it here. Bede Griffiths's language is telling. He writes of angels, of choirs of angels, of Paradise, and of God. The entire passage is imbued with the categories of a religion he had previously known, to which he assimilates his emotional response. As William James recognized so clearly, there are rival hypotheses about the sources of those feelings, potential psychological explanations in terms of very different causes. Secularism should acknowledge that experiences of this type currently lack full scientific explanation: it would be a mistake to assert, dogmatically, that there are antecedent features of Bede Griffiths's psychological state that would dispose him to feel just that surge of joy on just this occasion. We simply do not know what to make of some parts of our psychological lives, including these precious epiphanies. That does not mean, however, that we have a license to conceptualize them in terms of myths handed down to us from the remote past. Better to acknowledge our ignorance, to refrain from covering it with labels drawn from some unreliable tradition with which we are acquainted, and to look forward to some future possibility of a well-grounded explanation of these important forms of experience.

The remaining questions are harder, in large part because of well-known difficulties in comparing the experiences of different people and in assessing the effects that particular experiences have on subsequent conduct. The most obvious way in which to address question 3 is to ask those who have acquired or lost faith to consider the epiphanies they have had at different periods in their lives, and to compare them for depth and richness. As already noted, any such procedure is vulnerable to the objection that the judgments rendered are distorted by failures of memory and, possibly, later confabulation. It is easy to conjecture that those who have traveled in one direction (from faith to secularism, say) would give different verdicts from those who have made the opposite journey: those who have lost their faith declare that there has been no change in the quality of the experience; those who have acquired faith extol the gain in richness and depth. Judgments of this sort would, of course, cohere with the beliefs that the evaluators have at the time they make their assessments, and would embody a sense that they have now arrived at the correct view.

There is, however, a different way to approach the question, one in line with the discussions of section IV. What exactly could a sense of some higher, supernatural world add to these epiphanies? The obvious answer is that the religious person feels a connection to something far vaster than his or her own life. To accept that answer, however, would be to return to the issues of the previous section, specifically to the thought that seeing oneself as fulfilling an externally imposed divine purpose, even if it must be opaque to finite creatures like ourselves, is intrinsically more satisfying than anything secularists can offer. The attitude I attributed to secular humanism can be recapitulated here. Thoroughly secular people can interpret the purpose of their lives, not through some "vertical" links to a dimly understood transcendent reality, but through "horizontal" connections to a natural world that is vaster than their own individual existence. Recognition of yourself as part of a world, including most importantly other human lives, on which your actions make an impact, the epiphany can be a rich source of broader connections without any presuppositions about the supernatural. The religious claim of especial depth or richness in these experiences is thus exposed as the residue of misguided presuppositions that ought to be forsworn.

Questions 4 and 5 require more extensive concessions from the secularist. Despite the fact that many people who lack belief in the supernatural have had experiences through which they became committed to a course of action, a course they pursued with great diligence for the rest of their lives, I am prepared to allow that *as things currently stand*, the acceptance of a God may provide the epiphany with a force that pervades subsequent conduct. One of the most admirable people I know is a committed Christian, who, despite medical problems that have affected her for decades, is truly remarkable for the intensity and scope of her work in her community. She sustains and inspires her family, and simultaneously contributes extensively to the nurturing of the sick, the poor, and the aged. Her dedication is truly extraordinary, and it has often seemed incredible to me that, suffering as she does from multiple sclerosis, she manages to bring so much light into so many lives. Her own explanation for this is her recognition of God, once present in the person of Jesus—as she said to me when we last talked, "If I didn't believe that Christ really was who he said he was, I couldn't keep going with all these things."

Perhaps, then, the answer to question 4 is yes, and perhaps secularists should allow that, *in the social environments that currently exist*, there is an asymmetry between the motive force felt by those whose epiphanies are framed in religious terms and those for whom there is no sense of the supernatural. Similarly, it may be that the forms of the religious life make these important experiences more frequent or more enduring. Anyone sympathetic to the proposal that religions have evolved to meet human psychological and social needs ought to appreciate the possibility that the rituals and devotions in which religious people engage may have beneficial effects in directing human conduct and in providing a sense of "fullness." As section II constantly emphasized, the evolution of religious traditions doesn't have to accumulate or refine truths for the religion to succeed.

Inside or outside religious life, the effects of epiphanies upon us are often fleeting, and the sense of uplift offered by those experiences is hard to sustain. To Bede Griffiths's avowal, I juxtapose a (fictional) presentation by a far greater writer.

> He had confessed and God had pardoned him. His soul was made fair and holy once more, holy and happy. . . .He sat by the fire in the kitchen, not daring to speak for happiness. Till that moment he had not known how beautiful and peaceful life could be. The green square of paper pinned round the lamp cast down a tender shade. On the dresser was a plate of sausages and white puddings and on the shelf there were eggs. They would be for the breakfast in the morning after the communion in the college chapel. White pudding and eggs and sausages and cups of tea. How simple and beautiful was life after all. And life lay all before him.

The "he" of this passage is Stephen Dedalus, protagonist, if not "hero," of *Portrait of the Artist as a Young Man*, and the scene occurs just after Stephen returns from the confession to which the four sermons of the school retreat have driven him.[36]

Yet the uplift proves transient. Stephen quickly becomes caught up in a rigorous schedule of rituals that drain the vitality out of what he has experienced.

> Sunday was dedicated to the mystery of the Holy Trinity, Monday to the Holy Ghost, Tuesday to the Guardian Angels, Wednesday to Saint Joseph, Thursday to the Most Blessed Sacrament of the Altar, Friday to the Suffering Jesus, Saturday to the Blessed Virgin Mary. . . .His daily life was laid out in devotional areas. By means of ejaculations and prayers he stored up ungrudgingly for the souls in purgatory centuries of days and quarantines and years; yet the spiritual triumph which he felt in achieving with ease so many fabulous ages of canonical penances did not wholly reward his zeal of prayer since he could never know how much temporal punishment he had remitted by way of suffrage for the agonising souls: and, fearful lest in the midst of the purgatorial fire, which differed from the infernal only in that it was not everlasting, his penance might avail no more than a drop

of moisture, he drove his soul daily through an increasing circle of works of supererogation.[37]

Stephen is hardly bereft of epiphanies—indeed, the frequency with which he has them has made the term itself famous—but their effect on his life, whether in his devout or his lapsed phases, is typically fleeting. The significance of these episodes is very hard to maintain in one's consciousness, *and even the machinery of religion can erode any enduring effect.*

Joyce's point is, I think, a deep and important one. For almost all people, whether religious or secular, the occasions of uplift are rare, and their motive power is easily dissipated. That fact points toward the right way for secular humanism to respond to the challenge Taylor poses, as it emerges in the affirmative answers to questions 4 and 5 I have conceded. *Special forms of experience—epiphanies—are partly a social achievement.* The force that they have on human conduct, and the frequency with which they occur, depend on effective techniques that religions have introduced and honed over very long periods of time.

A commonplace about music asserts that the devil always has the best tunes. However that may be, when it comes to the cultivation of episodes that embody Taylor's "sense of fullness," religions have a long history of practice. They can draw, often brilliantly, on resonant words, forms of ceremony, art, and music, and the secular surrogates (for example, the "services" of Unitarians) frequently seem anemic by comparison. This means that the concessions I have made in response to questions 4 and 5 should be viewed not as marks of secularist loss, but as challenges to develop ways of sustaining those experiences we take to be most important that will be as powerful as those supplied by long-evolved religious traditions.

Dewey saw the point very clearly. Rather than suppose that epiphanies, like that of Bede Griffiths, owe their power to the belief in the supernatural, he suggested that we take the experiences for the valuable episodes they are, and find ways of sustaining and deepening them.

> It is the claim of religions that they effect this enduring change in attitude. I should like to turn the statement around and say that whenever this change takes place there is a definitely religious attitude. It is not religion that brings this about, but when it occurs, from whatever cause and by whatever means, there is a religious outlook and function.[38]

Decades later, secularism still needs to attend to the cultivation of this attitude, to elaborate ways in which it can become more widespread and more enduring. That task is plainly difficult, for the established religions of the world have honed their abilities to respond to the human need for community. There is, however, no reason to think that the obstacles are insuperable, or that secular humanists cannot find inspiration in those forms of religion that have committed themselves to nonliteral understandings of traditional doctrines. In outlining responses to some important challenges for secularism, I hope to have renewed the quest for what Dewey called a "common faith," a complex of psychological states beyond the acceptance of myth, that recognizes secular *human*ism as more than blunt denial.

# 2

# Disenchantment—Reenchantment

### Charles Taylor

These terms are often used together, the first designating one of the main features of the process we know as secularization, the second a supposed undoing of the first, which can be either desired or feared, according to one's point of view.

But their relation is more complicated than this. In some sense, it can be argued, the process of disenchantment is irreversible. The aspiration to reenchant (or the apprehended danger this threatens) points to a different process, which may indeed reproduce features analogous to the enchanted world, but does not in any simple sense restore it.

Let's speak of "the enchanted world" to designate those features which disenchantment did away with. There are two main ones.

The first feature of this world is that it was one filled with spirits and moral forces, and one, moreover, in which these forces impinged on human beings; that is, the boundary between the self and these forces was somewhat porous. There were spirits of the wood, or of the wilderness areas. There were objects with powers to wreak good or ill, such as relics (good) and love potions (not so unambiguously good). I speak of "moral" forces to mark this point, that the causality of certain physical objects was directed to good or ill. So a phial of water from Canterbury (which must contain some blood of the martyr Thomas à Beckett) could have a curative effect on any ill you were suffering from. In this it was quite unlike a modern medical drug that "targets" certain maladies and conditions, owing to its chemical constitution.

One could sum this up by saying that this was a world of "magic." This is implied in our term "disenchantment," which can be thought of as a process of removing the magic. This is even clearer in the original German: Weber's *Entzauberung* contains the word *Zauber* (magic). But this is less illuminating than it seems. The process of disenchantment, carried out first for religious reasons, consisted of delegitimizing all the practices for dealing with spirits and forces, because they allegedly either neglected the power of God or directly went against it. Rituals of this kind were supposed to have power of themselves, hence were blasphemous. All such rituals were put into a category of "magic." The category was constituted by the rejection, rather than providing a clear reason for the rejection. It then carries on in Western culture even after the decline of faith—for example, Frazer's distinction magic/religion. Only when Westerners attempted to make ethnographic studies of non-Western societies did it become clear how inadequate and instable this category is.

I talked about not being able to go back. But surely lots of our contemporaries are already "back" in this world. They believe in and practice certain rituals to restore health or give them success. The mentality survives, even if underground. That is true; much survives of the earlier epoch. But the big change, which would be hard to undo, is that which has replaced the porous selves of yore with what I would describe as "buffered" selves.

Let's look again at the enchanted world, the world of spirits, demons, moral forces that our predecessors acknowledged. The process of disenchantment is the disappearance of this world, and the substitution of what we live today: a world in which the only locus of thoughts, feelings, spiritual élan is what we call minds; the only minds in the cosmos are those of humans (grosso modo, with apologies to possible Martians or extraterrestrials); and minds are bounded, so that these thoughts, feelings, and so forth, are situated "within" them. What am I gesturing at with the expression "thoughts, feelings, and so forth"? I mean, of course, the perceptions we have, as well as the beliefs or propositions that we hold or entertain about the world and ourselves. But I also mean our responses, the significance, importance, meaning, we find in things. I want to use for these the generic term "meaning," even though there is in principle a danger of confusion with linguistic meaning. Here I'm using it in the sense in which we talk about "the meaning of life," or of a relationship as having great "meaning" for us.

Now the crucial difference between the mind-centered view and the enchanted world emerges when we look at meanings in this sense. On the former view meanings are "in the mind," in the sense that things have the meaning they do only in that they awaken a certain response in us, and this has to do with our nature as creatures who are thus capable of such responses, which means creatures with feelings, desires, aversions—that is, beings endowed with minds, in the broadest sense.

I must stress again that this is a way of understanding things which is prior to explication in different philosophical theories, materialist, idealist, monist, dualist. We can take a strict materialist view and hold that our responses are to be explained by the functions things have for us as organisms, and further by the kinds of neurophysiological responses that their perception triggers off. We are still explaining the meanings of things by our responses, and these responses are "within" us, in the sense that they depend on the way we have been "programmed" or "hardwired" inside.

The materialist fantasy, that we could, for all we know, be brains in a vat, being manipulated by some mad scientist, depends for its sense on this view that the material sufficient condition for thoughts of all kinds is within the cranium. Hence convincing thoughts about a nonexistent world could be produced through generation of the right brain states. The inside/outside geography, and the boundary dividing them, which is crucial to the mind-outlook, is reproduced in this materialist explication of it.

But in the enchanted world, meanings are not only in the mind in this sense, certainly not in the human mind. If we look at the lives of ordinary people—and even to a large degree of elites—five hundred years ago, we can see in myriad ways how this was so. First, as I said above, they lived in a world of spirits, both good and bad. The bad ones included Satan, of course, but beside him, the world was full of a host of demons, threatening from all sides: demons and spirits of the forest, and wilderness, but also those which can threaten us in our everyday lives.

Spirit agents were also numerous on the good side. Not just God, but also his saints, to whom one prayed, and whose shrines one visited in certain cases, in hopes of a cure, or in thanks for a cure already prayed for and granted, or for rescue from extreme danger, for example, at sea.

These extrahuman agencies are perhaps not so strange to us. They violate the second point of the modern outlook I mentioned above—that (as we ordinarily tend to believe) the only minds in the cosmos are humans—but they nevertheless seem to offer a picture of minds, somewhat like ours, in which meanings, in the form of benevolent or malevolent intent, can reside.

But seeing things this way understates the strangeness of the enchanted world. Thus precisely in this cult of the saints, we can see how the forces here were not all agents, subjectivities, who could decide to confer a favor. Power also resided in things.[1] For the curative action of saints was often linked to centers where their relics resided: either some piece of their body (supposedly), or some object that had been connected with them in life, like (in the case of Christ) pieces of the true cross, or the sweat-cloth that Saint Veronica had used to wipe his face, which was on display on certain occasions in Rome. And we can add to this other objects that had been endowed with sacramental power, like the Host, or candles that had been blessed at Candlemas, and the like. These objects were loci of spiritual power, which is why they had to be treated with care, and if abused could wreak terrible damage.

In fact, in the enchanted world, the line between personal agency and impersonal force was not at all clearly drawn. We see this again in the case of relics. The cures effected by them, or the curses laid on people who stole them or otherwise mishandled them, were seen both as emanating from them, as loci of power, and also as coming from the goodwill, or anger, of the saint they belonged to. Indeed, we can say that in this world, there is a whole gamut of forces, ranging from (to take the evil side for a moment) superagents like Satan himself, forever plotting to encompass our damnation, down to minor demons, like spirits of the wood, which are almost indistinguishable from the loci they inhabit, and ending in magic potions that bring sickness or death. This illustrates a point that I want to bring out here, and to which I will recur shortly, that the enchanted world, in contrast to our universe of buffered selves and "minds," shows a perplexing absence of certain boundaries that seem to us essential.

So in the premodern world, meanings do not inhere only in minds, but can reside in things, or in various kinds of extrahuman but intracosmic subjects. We can bring out the contrast with today in two dimensions, by looking at two kinds of powers that these things/subjects possess.

The first is the power to impose a certain meaning on us. Now in a sense, something like this happens today all the time, in that certain responses are involuntarily triggered in us by what happens in our world. Misfortunes befall us, and we are sad; great events befall and we rejoice. But the way in which things with power affected us in the enchanted world has no analogies in our understanding today.

For us, things in the world, those that are neither human beings nor expressions of human beings, are "outside" of mind. They may in their own way impinge on mind—really, in two possible ways.

We may observe these things, and therefore change our view of the world, or be stirred up in ways that we otherwise wouldn't be. Since we ourselves are bodies continuous with these external things, and in constant exchange with them, and since our mental condition is responsive causally to our bodily condition in a host of ways (something we are aware of without espousing any particular theory of what exactly causes what), our strength, moods, motivations, and the like, can be affected, and are continually being affected, by what happens outside.

But in all these cases, that these responses arise in us, that things take on these meanings, is a function of how we as minds, or organisms secreting minds, operate. By contrast, in the enchanted world, the meaning is already there in the object/agent: it is there quite independently of us; it would be there even if we didn't exist. And this means that the object/agent can communicate this meaning to us, impose it on us, in a third way, by bringing us as it were into its field of force. It can in this way even impose quite alien meanings on us, ones that we would not normally have, given our nature; as well as, in positive cases, strengthening our endogenous good responses.

In other words, the world doesn't just affect us by presenting us with certain states of affairs, which we react to from out of our own nature, or by bringing about some chemical-organic condition in us, which in virtue of the way we operate produces, say, euphoria or depression. In all these cases, the meaning  comes into existence only as the world impinges on the mind/organism. It is in this sense endogenous. But in the enchanted world, the meaning exists already outside of us, prior to contact: it can take us over; we can fall into its field of force. It comes on us from the outside.

The second feature of the earlier world that disenchantment sidelined is similar in import to the first. In another way, it placed meaning within the cosmos. Only this is a feature of elite culture. I am not speaking of popular "magic" and the sensibility of porous selves, but rather of high theory. The cosmos reflected and manifested a Great Chain of Being. Being itself existed on several levels, and the cosmos manifested this hierarchy, both in its overall structure and again in its different partial domains. The same superiority of dignity and rule as what the soul manifests over the body reappears in the state in the preeminence of the king, in the animal realm in that of the lion, among birds and fishes in the supreme status of eagle and dolphin. These features "correspond" to each other in the different domains. The whole is bound together by relations of hierarchical complementarity, which should be reproduced in a well-ordered state.

Once again, to point up the contrast with our world, we can say that in the enchanted world, charged things have a causal power that matches their incorporated meaning. The High Renaissance theory of the correspondences, which while more an elite than a popular belief, partakes of the same enchanted logic, is full of such causal links mediated by meaning. Why does mercury cure venereal disease? Because this disease is contracted in the market, and Hermes is the god of markets. This way of thinking is totally different from our post-Galilean, mind-centered disenchantment. If thoughts and meanings are only in minds, then there can be no "charged" objects, and the causal relations between things cannot be in any way dependent on their meanings, which must be projected on them from our minds. In other words, the physical world, outside the mind, must proceed by causal laws that in no way turn on the moral meanings things have for us.

We can see how elite theory and popular sensibility interpenetrated and strengthened each other. The high theory was easier to believe in a world of enchanted sensibility. And the theory itself could draw on some features of popular lore, giving them a new rationale and systematic form.

This second feature is easier to imagine recovering in our world. Certainly many people hold "wacky" theories. But a wholesale acceptance of this outlook as a hegemonic one is surely unthinkable in the post-Galilean world.

## 2

Now what do people seek who look to "reenchantment"? In a sense, it is the same fundamental feature, but differently conceived. In other words, they bridle at the idea that the universe in which we find ourselves is totally devoid of human meaning. Of course, instrumental meaning can be attributed to various features of our natural surroundings, in virtue of their serving or impeding our organic needs, but any human meaning must be simply a subjective projection. By "human meaning," I mean what we try to define when we identify the ends of life, through judgments such as these: this is really meaningful as a way of life; or this life is really worth living; or this form of being is a real fulfilment, or a higher way of being, and the like. Derivatively, we can attribute human meaning to the things that surround us because of their role in these ends or purposes. A statement like Thoreau's "in wildness is the preservation of the world" is such an attribution of meaning. It is the kind of statement that proponents of reenchantment often want to make.

This sense of loss was frequently expressed in the Romantic era. Take Schiller's poem "The Gods of Greece."

> Da der Dichtung zauberische Hülle
> Sich noch lieblich um die Wahrheit wand,
> Durch die Schöpfung floss da Lebensfülle,

Und was nie empfinden wird, empfand.
An der Liebe Busen sie zu drücken,
Gab man höhern Adel der Natur,
Alles wies den eingeweihten Blicken,
Alles eines Gottes Spur.

(When poetry's magic cloak
Still with delight enfolded truth
Life's fulness flowed through creation
And there felt what never more will feel.
Man acknowledged a higher nobility in Nature
To press her to love's breast;
Everything to the initiate's eye
Showed the trace of a God.)

But this communion has now been destroyed; we face a "God-shorn nature":

Unbewusst der Freuden die sie schenket,
Nie entzückt von ihrer Herrlichkeit,
Nie gewahr des Geistes, der sie lenket,
Sel'ger nie durch meine Seligkeit,
Fühllos selbst für ihres Künstlers Ehre,
Gleich dem toten Schlag der Pendeluhr,
Dient sie knechtisch dem Gesetzt der Schwere,
Die entgötterte Natur.

(Unconscious of the joys she dispenses
Never enraptured by her own magnificence
Never aware of the spirit which guides her
Never more blessed through my blessedness
Insensible of her maker's glory
Like the dead stroke of the pendulum
She slavishly obeys the law of gravity,
A Nature shorn of the divine.)

And so what seems wrong with total disenchantment? What makes people seek reenchantment? Now the complaint that one finds again and again, in what I will call loosely the post-"Romantic" period, targets a reading of our modern condition in which all human meanings are simply projected. That is, they are seen as arbitrarily conferred by human subjects. None would be valid universally. Universal agreement on these meanings would result from de facto convergence of our projections.

Thoreau's statement about wilderness would have to be read as one such subjective projection, rather than claiming validity for all human beings.

But this projectivist outlook doesn't follow from disenchantment in the double sense outlined above. True, human meanings are no longer seen as residing in the object, even in the absence of human agents. These meanings arise for us as agents-in-the-world. But it doesn't follow from this that they are arbitrarily conferred.

There is a massive slippage in the reasoning here, which has frequently accompanied the modern turn to the subject. In the field of epistemology, this turn (Descartes, Locke) first of all generated a view of knowledge as a correct portrayal of external reality residing *in* the mind. But this reflexive turn to examine our experience, carried through more fully, ended up dispelling this illusion. Our grasp of the world is not simply a representation within us. It resides rather in our dealing with reality. We are being in the world (Heidegger's *Inderweltsein*), or being to the world (Merleau-Ponty's *être au monde*).

Some similar working through needs to be done in this domain of human meanings. Otherwise we are living with a distorted view of ourselves.

A word about the relation of this debate to that between theists and atheists. Obviously the former will not consent to the notion that there is no meaning in reality in abstraction from human agents; and I myself am not accepting this. But this doesn't make it any the less interesting for me to inquire how we can discover in human experience what meanings must be recognized as universally valid. The debate between theists and atheists can be better conducted once we have established what these meanings are. Without taking this first step, we are living a distorted form of the human condition, where instrumentalist deviations can put in danger the very survival of the planet.

So the issue about reenchantment can be put this way: when we have left the "enchanted" world of spirits, and no longer believe in the Great Chain, what sense can we make of the notion that nature or the universe which surrounds us is the locus of human meanings which are "objective," in the sense that they are not just arbitrarily projected through choice or contingent desire?

Put another way, attributions of these meanings count for us as strong evaluations. The distinction between strong and weak evaluation that I'm adverting to here, comes to this. A weak evaluation is one depending on choices that we may not make, or on our espousing ends that we may not accept. We can thus defeat the claim that something should have value for us, by choosing another end, or repudiating the one on which this value depends. In the case of strong evaluations, we cannot so release ourselves, and our attempt to do so reflects negatively on us.

This distinction for the moral realm can be illustrated by Kant's contrast between categorical and hypothetical imperatives. If someone says: invest in real estate (admittedly not a smart idea today, but often sound advice), you can defeat the imperative by saying: I'm rich enough already; or I'm dedicated to a life of poverty. But if someone says: act to reduce unnecessary suffering, you couldn't release yourself by pleading that you have other goals in life. You would need an argument to the effect that reducing suffering wasn't good, because, for example, it leads to a world of "last men," or blocks the way to the *Übermensch*.

But, of course, strong evaluations can be made outside the moral realm, in aesthetics, for instance. A judgment that some work or scene is beautiful would be strong if it carried the implication that those who could not concur were defective in their perception of beauty, rather than just being uninterested in this kind of thing.

The attribution of human meaning to (things in) the universe as a strong evaluation straddles the gap between the moral and the aesthetic. It concerns perhaps the ethical in the broad sense, where we make judgments about what a really good or properly human life consists in.

With these reflections as background, let's try to look at the debates about disenchantment and reenchantment. I follow here the excellent discussion by George Levine in his recent book.[2] The debate starts from the claim made by many people, both atheists and theists and some people in between, that the combination of Weberian rationalization and post-Galilean science, with the accompanying decline of religion, has left us with a world deprived of meaning, and offering no consolation. The situation of moderns is thought to be very different in this regard from that of people in all previous ages and cultures. A debate may then break out over what we can or ought to do about this: face the empty world with resolute courage, or call into question the rejection of religion, or perhaps find some new sources of meaning.

But one might instead interrogate this "Weberian" picture.[3] Is this really our predicament? Some have called this into question. Do we not now experience a wonder at the vast yet intricate universe, and the manifold forms of life, at the very spectacle of the evolution of higher forms out of lower ones? Do we not find beauty in all this? In this case, a part of the very change that is held to have disenchanted the world, here that bit of modern science which we call the theory of evolution, has in fact given us further, deeper cause to wonder at the universe. As Levine puts it, the world hasn't lost meaning; "it is stunning, beautiful, scary, fascinating, dangerous, seductive, real."[4] The first four of these epithets might be thought to be rather aesthetic than ethical. But one can argue that this sense of greatness and beauty fosters a love of the world which is one of the wellsprings of generosity. As Kant saw, the inspiration we draw from "the starry skies above" is akin to that we sense before "the moral law within."[5]

In fact, from the very beginning, materialism has generated a sense of awe at the universe and at our genesis out of it. One can find this in Lucretius, but this sense intensifies and develops in the eighteenth century, as the materialist outlook takes on shape and consistency. This vision solidified, but it also deepened.

We are alone in the universe, and this is frightening; but it can also be exhilarating. There is a certain joy in solitude, particularly for the buffered identity. The thrill at being alone is part a sense of freedom, part the intense poignancy of this fragile moment, the *dies* (day) that you must *carpere* (seize). All meaning is here, in this small speck. Pascal got at some of this with his image of the human being as a thinking reed.

The new cosmic imaginary adds a further dimension to this. Having come to sense how vast the universe is in time and space, how deep its microconstitution goes into the infinitesimal, and feeling thus both our insignificance and our fragility, we also see what a remarkable thing it is that out of this immense, purposeless machine, life and then feeling, imagination, and thought emerge.

Here is where a religious person will easily confess a sense of mystery. Materialists often want to repudiate this; science in its progress recognizes no mysteries, only temporary puzzles. But, nevertheless, the sense that our thinking, feeling life plunges its roots into a system of such unimaginable depths, that consciousness can emerge out of this, fills them, too, with awe.

Our wonder at our dark genesis, and the conflict we can feel around it, is well captured by a writer of our day. Douglas Hofstadter recognizes that certain people

> have an instinctive horror of any "explaining away" of the soul. I don't know why some people have this horror while others, like me, find in reductionism the ultimate religion. Perhaps my lifelong training in physics and science in general has given me a deep awe at seeing how the most substantial and familiar of objects or experiences fades away, as one

approaches the infinitesimal scale, into an eerily insubstantial ether, a myriad of ephemeral swirling vortices of nearly incomprehensible mathematical activity. This in me evokes a cosmic awe. To me, reductionism doesn't "explain away"; rather, it adds mystery.[6]

But this awe is modulated, and intensified, by a sense of kinship, of belonging integrally to these depths. And this allows us to recapture the sense of connection and solidarity with all existence that arose in the eighteenth century out of our sense of dark genesis, but now with an incomparably greater sense of the width and profundity of its reach.[7]

And so materialism has become deeper, richer, but also more varied in its forms, as protagonists take different stands to the complex facets I have just been trying to lay out. The reasons to opt for unbelief go beyond our judgments about religion, and the supposed deliverances of "science." They include also the moral meanings that we now find in the universe and our genesis out of it. Materialism is now nourished by certain ways of living in, and further developing, our cosmic imaginary; certain ways of inflecting our sense of the purposelessness of this vast universe, our awe at, and sense of kinship, with it.

But one cannot just stop here. A hostile critic will object that these are merely *feelings*. We sense depth and greatness, but does this correspond to reality? Is there a reality to which these feelings can correspond, if we do away with all religion and metaphysics? To which one can reply in turn that just having discovered a new and (supposedly) more correct explanation of things shouldn't alter in any way the awe or admiration we feel before these things. Thus, in this case, where we explain the shape of the universe, or the origins of life, we can split completely (a) our account of the origins of things, from (b) our sense of the significance they bear.

Now there is some truth in this reply, but the matter is more complicated than it implies. When we talk of our sense of wonder at the greatness and complexity of the universe, or of the love of the world it inspires in us, these are what I called above strong evaluations. They carry the sense that wonder is what one should feel; that someone who fails to sense this is missing something, is somehow insensitive to an object that really commands admiration. In this it is quite unlike one's preference, say, for flavors of ice cream, to take a really trivial and hence clear contrast case. I like strawberry and you vanilla. It would be absurd for us to accuse each other of not seeing something.

Let's take up this point with another contrast. The understanding behind strong evaluations is that they track some reality. A question can be raised: is that object really worthy of respect, or of wonder? That object may inspire love in you, but does it merit love? Obviously these arguments are central to ethical life. Yes, one should do the courageous thing, but is that (blindly charging the enemy) really courage? Be generous, but is that (spending from the PR budget to raise the profile of the company) really generosity? Now take a (rather shocking) contrast: we feel nausea at certain things; we can call these "nauseating." Does it make sense to argue about what is *really* nauseating? I am thinking of the literal applications to substances whose sight or smell is unbearable; of course there are moral applications, as when I say that a government policy is nauseating. Clearly the answer is no in the straightforward, literal cases. What nauseates us is a brute fact; some things just trigger this response, others not. If there are interpersonal differences, they also are a matter of brute fact. There is no arbitrating them.

These contrasts point up the fact that underlying strong evaluations there is supposed to be a truth of the matter. And this can't be separated from facts about how our reactions are to be explained. Put simply, our moral reactions suppose that they are responses to some reality and can be criticized for misapprehension of this reality. But in the case of nausea, there is no room for this account; the issue of reality can't get a purchase.

Now if we keep this distinction in mind, we might agree that there is indeed no reason why switching from a theistic to a materialist account of reality should undercut our wonder at the universe, although the account of what properly inspires wonder will be different, and will connect to different things for the theist and atheist, respectively.

But take something like a sociobiological account which claims to show that human behavior, like all animal behavior, is really driven by "selfish" genes. There is, indeed, an "altruist" pattern of action, where one agent benefits another to his own detriment; but there is no real difference in the underlying motivation. This has to undercut part of the crucial background understanding of those who admire altruism. Why? Because admiring altruism is not just finding this pattern of action useful; it is also holding that the motivation which powers it is in some ways higher, more noble, more admirable. A claim like this can make sense only against a background view of human motivation as capable of transformation so as to be more and more drawn by the higher, so that ultimately our pattern of action is changed. The background is also richer than a simple hierarchy of motives; there is invariably some account of why one motive is higher: because it brings true harmony, corresponds to our real self, brings about a unity and harmony between human beings that answers one of our deepest longings, and so on through a wide range of alternatives. It is this whole background outlook that is negated and made impossible by the claim that we are always and inalterably at base "egoistic."

One can see a parallel case in one of Nietzsche's lines of attack against Christianity. Christians speak of "charity," of "love," of turning the other cheek. But in fact this is a lie. Because really we are all actuated by the will to power. Christian behavior is in fact motivated by a desire to get back at, and perhaps even get control over, those who have bested them in the power stakes. Now from the Christian point of view, going the extra mile, pushing further in self-giving love, makes sense, because it can be a step in a transformation that makes us more Christlike, more God-like. If this entire prospect is a delusion, if we are all inalterably and equally actuated by the Will to Power, then the aspiration is vain; the whole outlook collapses. It is just an extra ironic twist in Nietzsche's argument that he shows that "love" here is really driven by its opposite, *ressentiment* and hatred.

Thus, while it is certainly right to say that disenchantment in the double sense above, and a rejection of religion, can't by themselves undercut very powerful human meanings, like a sense of wonder, love of the world and of others, and the like, it is clear that certain kinds of reductive accounts of human life and action do rule these out. We can't just say that explanations of why we experience these meanings are irrelevant to their validity; that they stand on their own, because we *feel* them strongly. Our attributing these meanings makes a stronger claim. It lies in their nature as strong evaluations to claim truth, reality, or objective rightness.

That is why those who experience the wonder at reality that I described above feel the need to make sense of this, to articulate it. We saw above that a whole background understanding underlies the different ethics of altruism. This background will be different for theists and for atheists, but neither can rest simply on the bare reaction. One could argue that in human life such reactions are primary, and that all attempts to develop philosophical and religious theories flow from the need to articulate them. But these articulations are not simply derivative and secondary. Each way of giving expression to what appears as a similar reaction modifies it, develops it, gives it a different thrust. Thus we see how today in the sphere of humanitarian work, people of all convictions, religious and nonreligious, work side by side. They are in a sense actuated by the same impulse, but some very different ethics lie behind the common dedication in each case: different views of human life, of the possibilities of transformation, of the modes of spiritual or mental discipline that are to be engaged in, and so on.

But these articulations are never complete. They leave gaps where mystery intrudes, where the claims to truth are not fully grounded, where seeming refutation or contradictions lie half-visible. In a sense, all require some degree of faith that these difficulties could in principle be resolved. And this is true as much of the naturalist and materialist ones as of the "faith-based" ones. Indeed, though the gaps and the uncertainties lie in different places, this term could apply to all humanitarian efforts, however motivated.

Some might be astonished that something like a truth claim can ever be made in the moral or ethical domain. How can one ever prove conclusively that some direction of human transformation, toward sainthood, or Buddhahood, or a pure Kantian will, can really be accomplished, that there aren't decisive obstacles on the way? Indeed, this cannot be shown in the classical "scientific" way, that one can find some reality which corresponds to and thus verifies the hope. Some (the companions of the Buddha, the disciples of Christ) believed themselves to have seen such a reality, but this cannot simply be manifested to others who have not. In fact, a good deal of our confidence in our own faiths comes through another route, the unmasking of our illusions that allows us to abandon earlier, more questionable views. This process is, of course, never complete, but views that can survive such winnowing of illusion are less incredible than those that cannot.[8]

# 3

The above discussion can perhaps help to define better the question: in what way does a scientific account of the world "disenchant" it beyond recall, that is, beyond any possibility of what we have been calling reenchantment? Clearly the simple fact that we understand better how different species evolved, however "mechanistic" the process identified, cannot take away from our wonder at the scope and intricacy of the resultant system. We are faced, however we understand it, with the fact that we can thus respond with wonder; we might want to add that anyone with sufficient knowledge, training, consciousness cannot but feel this wonder.

Where, then, can the potential conflict lie? It lies in the fact that this wonder is lived by us as a strong evaluation; one that in some ways ennobles our lives, a wonder whose absence betokens a lack in the person left indifferent by these marvels. Now like all strong evaluations by which we give sense to and live our lives, this would run athwart a certain kind of reductive explanation, not of the evolution of birds on Pacific islands, but of our own psychology and behavior.

Wherein does this conflict reside? A response that we understand as a strong evaluation supposes the following ontology: (a) This response genuinely motivates us—it is not simply a cover, or a rationalization, or a screen for some other drive; (b) it can fail to occur on some occasions or in some people, but this betokens some limitation, blindness, or insensitivity on their part; (c) in other words, there is something objectively right about this response; (d) we can and ought to set ourselves to cultivate this response, refine/improve our perception of its proper objects. This four-point feature represents a package, reflecting our sense that this evaluation is founded. In Bernard Williams's terms, our moral and other strong evaluations claim to be "world-guided."[9]

Now an account in blind, mechanistic terms of the evolution of the birds poses no essential threat to my wonder—although a conflict may in fact arise here if my sense of awe is bound up with a belief in conscious design. But an account of my psychology and action, and hence of the production of this response on certain occasions, that admitted only "blind," efficient causes does create a conflict, as we saw with the case of altruistic motivation above. Here is where the issue of reduction crucially enters.

The question of reductive explanation arises when we have two accounts of what would seem to be the same range of phenomena (where "same" can be glossed in terms of spatiotemporal coordinates, or other uncontroversial ways of identifying the same objects). Now accounts couched in terms of post-Galilean science clearly are intended to avoid teleology or intentionality, purpose or evaluation, as causally relevant factors, while accounts of what we are doing that recognize strong evaluation make essential reference to such factors. The issue about reductive explanation is this: can we give an adequate explanation of the phenomena we describe in the "upper" language (e.g., people reacting to the universe with a sense of wonder) entirely in the terms of the "lower" language (post-Galilean science)? This would mean, inter alia, that we could provide necessary and sufficient conditions for all states described in the upper language in the terms of the lower language. But then the claim, made on behalf of the "upper" phenomena, that they operate on genuinely different principles from those adequately explained by the lower language would be voided.

Of course, some philosophers would want to deny any difference in ontology between humans and machines. The only difference between humans, on one hand, and machines that seem to exhibit purposive behavior, like guided missiles and computers, apart from a difference of complexity, lies in the stance that we assume to them. As observers we can adopt the "intentional" stance to both human and missile, or we can see them as mechanisms, which is the more fruitful language for explaining what they do, but both are explained by the same principles. Could we, then, attribute to more complex robots consciousness and self-feeling? Why not? argues Daniel Dennett.[10] Perhaps the robot feels that it is striving to do its task conscientiously. But apart from the difficulty of attributing consciousness and feeling to beings made of plastic and silicone, this inner sense would be in the robot purely epiphenomenal, perhaps itself built in an extra feat of engineering, but not essential to the account of what it is doing.

But our whole moral-evaluative lives rest on the opposite understanding. Evaluations motivate, and they reflect a perception of their objects that can be more or less adequate.

It is in fact extraordinary how blithely some philosophers can talk of reductive explanation as a virtually certain prospect, given all that we know about human life and evolution. A neurobiological account of human conscious action, in terms of how we are "wired," would operate primarily on the level of the individual organism and would explain social combinations in terms of the interactions between such organisms. But human culture is in fact developed in societies. As Merlin Donald has pointed out, no human being invents language, much less human culture, on its own. Left to themselves, abandoned children, or those who are unreachable, like Helen Keller, never take the step to language and the conscious grasp of things that language enables.[11] What is more, the "wiring" of the brains of young infants is to a significant extent determined in the course of their early upbringing by the kinds of things they learn, which themselves vary from culture to culture. No one can simply deny a priori that all this can be captured in terms of a monological, purely efficient causal account of the formation and operation of brain and nervous system, but it is not easy to see how this can be accomplished.

On top of this, we have the fact that cultural evolution has brought about new strong evaluations, such as those we live by today, like democracy and universal human rights, and we have quite an agenda for anyone who would give a more circumstantial account of the evolution of humankind. This is a tremendously valuable goal, but it is not immediately obvious how we should go about realizing it.

It follows from the above that the issues of disenchantment and reenchantment arise on at least two levels. First, there is the question of whether disenchantment in the senses I described at the outset—that is, the dissipation of the enchanted world, the denial of the Great Chain, plus the widespread rejection of Western theism—have not voided the universe of any human meaning. In particular, it has been claimed here that there is no further basis for a sense of awe and wonder at the universe, which in turn can inspire in human beings love and even gratitude toward the greater whole in which they are set. The answer to this is that undoubtedly some modes of wonder, articulated in a certain manner, will be decisively undercut. But the question remains open whether other forms, based on our own experience of being in the world, can be recovered. It seems to me that the answer here is affirmative.

But there is a quite different way in which the deliverances of "science" could undercut this affirmative response, and that is by a certain form of reductive explanation, not of the universe, but of human life. This is perhaps one of the most burning intellectual issues in modern culture.

Of course, to revert to the first issue, there remains a question whether purely anthropocentric articulations can do justice to our sense of wonder, and other related evaluations. This will remain a bone of contention between people in different positions, religious, secular, spiritual, indeed of an almost unlimited variety. The discussion between us promises to be fruitful; there is a virtual infinity of insights here, of which no single view has the monopoly. But all of these depend on a rejection of a reductive account of human life

# 3

## ↭

# Enchantment? No, Thank You!

### *Bruce Robbins*

The eminent translators of Max Weber's great lecture "Science as a Vocation" (1917/1919) put the phrase "disenchantment of the world" in quotation marks.[1] This may be because, as they note in their introduction, Weber was borrowing the phrase from Schiller. But the German text has no quotation marks.[2] This is a small thing and perhaps a trivial one. Still, it hints that Gerth and Mills may have been putting a certain distance between Weber and the historical process with which his name has been so tightly associated. Perhaps they merely wished Weber had taken such a distance. On the other hand, perhaps they had decided, on the basis of what they knew, that he could not have given himself over fully to the disenchantment story. The latter hypothesis seems quite plausible. As it turns out, there is evidence that Weber was not merely ambivalent about disenchantment—that is the standard view—but that he did not believe the disenchantment of the world had happened, at least as it has been popularly understood. I would like to pursue this idea here, not so much for its pertinence to the understanding of Weber (I leave that to the Weber scholars) as for its own sake. Whether or not Weber did believe in it, disenchantment seems to me one of the more disabling and sneakily misleading stories we are in the habit of telling ourselves regularly. It's a habit I'd like to see us kick.

The outline of the story is simple: there is something called enchantment that (1) we once had, but (2) we have since lost, and (3) we are now in dire need of. Opinions differ as to whether it can be retrieved, but everyone agrees that what has been lost is extremely valuable. Each of these propositions is problematic in its own right. Let us begin with the valuable something called enchantment.

Enchantment sounds like religion, but Weber seems to have been at least moderately eager to distinguish the two. Schiller's version of disenchantment—it turns out that Weber was not quoting him directly—is

*Entgötterung* (de-divinization). When Schiller used the phrase "die entgötterte Natur" (nature from which the gods have been eliminated) in
his 1788 poem "The Gods of Greece," he was criticized for seeming to
lament the end of polytheism, and he backed down.[3] Weber's term is
*Entzauberung* (the elimination of magic). It may bow gently to Schiller
but, whether for reasons of diplomacy or not, it certainly takes the emphasis off divinity. In *The Protestant Ethic and the Spirit of Capitalism*
and elsewhere, Weber had already argued that religion was responsible
for the elimination of magic; Christian monotheism could not be seen as
the simple victim of the disenchantment or rationalization of the world,
since it was (also) an agent of that process, helping to thin out the pagan
population of spirits and demons. In any case, the switch from gods to
magic makes Weber's version sound more inclusive. Orthodox belief is
not the object Weber is chiefly mourning. Whatever magic is a figure for,
nonbelievers suffer from its loss as much as do believers. That at least
seems to be the popular understanding.

But if enchantment cannot be simply identified with religion or tradition, then how *should* it be understood? According to Jane Bennett,
"*enchant* is linked to the French verb *to sing: chanter*. To 'en-chant': to
surround with song or incantation; hence, to cast a spell with sounds,
to make fall under the sway of a magical refrain, to carry away on a
sonorous stream."[4] Here being enchanted sounds like an innocent good
time, entirely lacking in those unbending dogmas and rigorous obliga-
tions that burden tradition and religion alike. There is no obvious reason
to resist it. But before you are swept away by Bennett's mellifluous poem
in prose, pause to consider that the words "cast a spell" and "fall under
the sway" are at least as crucial to the history of the term as the sonorous stream of song. What a modern English speaker hears in the word
"enchantment" is the state of being exceptionally charmed, delighted,
enraptured, as by an encounter with a person or an artwork. But all this
is of course dead metaphor, and it needs to be dug up, cleaned off, and
seen for what it is. Once upon a time—that is, in the time before the
world was disenchanted—enchantment would have meant, literally, the
employment of magic or sorcery. It presumed a situation in which people
considered themselves in constant threat of being attacked and having
their will overridden by the irresistible compulsion of a spell. The disenchantment story borrows its movie-blurb vocabulary ("enchanting!"
"spellbinding!") from a sense of causality that would have involved serious psychic inconveniences for those concerned. Conservative Christians
still inhabit that threatening mental universe—that's why, unlike the rest

of us, they cannot safely enjoy J. K. Rowling's Harry Potter books as fantasies (which is a shame, since in many ways they are quite conservative fantasies).[5] What the disenchantment story does is invite the rest of us to abandon ourselves to the desire for figurative enchantment without worrying what it would be like actually to believe we are vulnerable to spells, that the world is full of magicians and malevolent spirits that at any moment could harm us, strike our loved ones with paralysis or disease, destroy our crops and livestock, and so on.

The point is not incidental. As Charles Taylor argues in *A Secular Age*, a rereading of the disenchantment narrative that I will discuss in some detail below, for most of the history of the West it was the certainty that evil spirits existed that made disbelief so very difficult for most people even to conceive of, let alone enter into. You needed Christianity's good spirits because you needed protection against the evil spirits who, you had no doubt, lurked in wait all around you. Nature back then was famously animated, and the spirits were what animated it—animated it with intentions, but not necessarily good intentions. The disenchantment story celebrates the premodern past without bothering to remember the evil spirits. The sort of belief it tells us we need is defined by this elision of the bad stuff from back then. We are urged to buy the whole package without being reminded of half of its contents.

There are of course versions of the disenchantment story that don't seem to be a story at all. The concept is at its weakest when the crudeness of "before" and "after" is most exposed. This is especially true for the embarrassing "before": when you can't avoid identifying enchantment with, say, the notion that you can successfully propitiate the gods and get favorable winds to blow by sacrificing your daughter. But it also holds for the logic that links them: the highly improbable assumption that what society today is most sorely missing is (of all things) precisely and symmetrically what society used to have. Disenchantment seems a more plausible proposition when it cuts directly to the present, insisting on how deeply desirable enchantment is whether or not it is an aspect of the past that can now be retrieved. In Bennett's version, for example, enchantment is more or less liberated from the encumbrance of the disenchantment plot. Bennett argues quite persuasively against that plot. She breaks the narrative apart by dismissing stage 2, enchantment lost. And, unlike Taylor, she pays little attention to the idyllic stage 1, the premodern era when enchantment supposedly reigned.[6] But her lack of interest in stage 1 and her dismissal of stage 2 in no way diminish her commitment to stage 3, the present desirability of enchantment, which remains perfectly intact. Bennett never stops to question the idea that enchantment is what we want and what we need. On the contrary, she allows the concept to expand imperiously. Enchantment as she presents it is a sort of magical solution to a great deal of what ails us, ready to sort out our most pressing ethical and political dilemmas while also realigning our emotions and even our bodily existence. Luckily it's right here, fully available, all around us.

The present volume doesn't go quite so far. Still, the paradox written into the idea of "secular enchantment" seems again to want to have it both ways, narratively speaking. Disenchantment has happened, so the book implies: that's why the world is secular, why we have the specific problems (no natural basis for ethics, no meanings or values that reach beyond self-satisfaction, no compensation for life's inevitable pains, losses, and injustices) for which enchantment can be proposed as the solution. But disenchantment also cannot have happened, or else enchantment would not be still so readily available. This is probably not too far from what I believe myself, with some readjustment of emphasis. (To put this alternative in the same paradoxical form: disenchantment has not happened, and therefore enchantment is not the name of what we need. And disenchantment has happened, but what has been produced thereby is secularism, which is less of a problem than a solution, though by no means a final or inclusive one.) But I would prefer, if possible, to put my beliefs in terms that are less paradoxical. I suggest, then, that we examine the paradox at both ends. Are we sure it is indeed enchantment, and not some other thing entirely, that we really need and want? And has there occurred something that can properly be called disenchantment?

೧೨

There is no doubt that "the enchanted world" was "an object of longing" for Weber, that he saw enchantment through a "haze of nostalgia." Still, nostalgia was only one of Weber's inspirations, and what Bennett calls "the problem of meaninglessness" was by no means the key result of disenchantment as Weber imagined it.[7] Here a bit of social context is helpful. As Fritz Ringer points out in *The Decline of the German Mandarins*, the mandarin elite to which Weber belonged was "rapidly losing its influence upon the new electoral politics." For this reason and others, it tended to be pessimistic about democracy and quick to confuse democratic institutions or reforms with the supposed soullessness of a bleak, alienated, meaning-deprived modernity. Weber shared this nostalgic sensibility. But he also stood fast against the "thoughtlessly unqualified rejection of industrial civilization itself" by which many of his fellow mandarins were attracted (as many champions of disenchantment are today). "Weber himself tended to speak of rationalization and bureaucratization as more or less unpleasant inevitabilities. His tone was that of the heroic pessimist who faces facts; but it was also characteristic of him that he would not tolerate the obscurantist illusion of a total escape from modernity."[8]

If Weber did not fall for or call for that illusory escape from modernity, it was not simply because (as those here taxed with obscurantism would perhaps reply) he had fallen instead for the modernist myth of the social scientist or scholar (*Wissenschaft* in the lecture's title is better translated "scholarship" than "science") as a new charismatic hero, facing alone the meaningless void that the fearful masses could not be expected to inhabit or even acknowledge. That myth had (and has) its attractions. But it is counterbalanced in Weber by elements of a history that does not, as later readers have assumed, begin in meaning and end in meaninglessness. According to Catherine Colliot-Thélène, there are two basic reasons why Weber should not be read as proposing, even ambivalently, a nostalgic narrative of rationalization. First, the "after" of Weber's narrative was not simply rational. Second, the "before" was not simply irrational.

Weber sees religion as both an agent of rationalization, eliminating magic from the world, and as itself in a real sense quite rational. What people want and have always wanted from religion, in Weber's eyes, is not so much personal alignment with ultimate cosmic truths as material, this-worldly things like good health, prosperity, and progeny. Religion is not alien to instrumental rationality. (That is also Taylor's point about protection against evil spirits, though he later forgets he's made it.) When Weber describes religion, then, he does not put the emphasis on "the beyond," as if this-worldly life were inherently inadequate without some reference to a mysterious or unknowable higher reality. On the other hand, the iron cage of modernity could be described equally well as hyperrational or as just the opposite.[9] At the end of *The Protestant Ethic and the Spirit of Capitalism* Weber paints a portrait of unbridled, endless acquisitiveness, shaken loose from the rational-religious goal of increasing the capitalist's points in heaven or even his physical well-being, unable to remember why, though rich, he is still exhausting himself to pile up further wealth. This portrait is not endearing, but it cannot count as proof that what modernity has done is rationalize the world. The same holds for Weber's insistence that despite the process of disenchantment the modern world remains, at least metaphorically, polytheistic. More on this below.

As Colliot-Thélène demonstrates, Weber's various mentions of disenchantment do not all refer to the same thing. "In different passages," she comments, "Weber gives his famous formula either the limited, precise sense of the elimination of magical techniques for salvation, or the wider sense of the abandonment of the quest for meaning."[10] What readers have done is to jump from *magic*, which is historically specific and easier to give up on, to *meaning*, which seems universal and absolutely indispensable. The loss of the first therefore seems to entail the loss of the second. This is how the story of disenchantment is most often read. For the popular understanding, religion and its near equivalents exist in order to satisfy a hunger for meaning, and this hunger is universal and ineradicable. To be without meaning, on this view, is necessarily to fall into a famished and pathological state, like Durkheim's anomie. It's the ability to fill up anomie's presumed emptiness that explains the desirable social effects attributed to religion. Hence enchantment—religion plus its near equivalents—is something that society needs, and will always need.

And perhaps, if society has not yet starved to death, meaning is something that society will always already have supplied. Here the story collides with itself, implying that, if the need is indeed so peremptory, meaning cannot really ever have been absent. To say that disenchantment *means* the loss of meaning, making use of the predicate as a verb in the very act of declaring that it is lost, is both to risk self-contradiction and to make that loss seem a lesser, much more uncertain thing. Reading the disenchantment story in this way gives it a flatter arc, without either the high peaks of the enchanted before and the reenchanted after or the deep, deep valley of disenchantment in the middle. To go a bit further in this direction is to suggest that there may not be a narrative here at all. Perhaps meaning is something that one simply cannot not have. In *The Meaning of Life*, Terry Eagleton distinguishes between two senses of meaninglessness: "when somebody wails 'My life is meaningless,' they do not mean that it makes no sense in the way that '&$£%' makes no sense. Rather, it is meaningless more in the sense that 'Assuring you most earnestly of our respectful attention at all times, we remain your obliged and most devoted servants . . .' is meaningless. People who find life meaningless are not complaining that they cannot tell what kind of stuff their body is made out of. . . . They mean, rather, that their lives lack *significance*. And to lack significance means to lack point, substance, purpose, quality, value, and direction."[11] Significance in this sense does not require any transcendental grounding. Even "Assuring you most earnestly of our respectful attention at all times" has a legible social purpose behind it, though not that of communicating ceaseless individual attention or respect. The world is full of social purposes. The question is which ones you like and which ones you don't.

Weber's philosophical point is, I hope, uncontroversial: that human-kind must learn to make do without natural or transcendental foundations for its ethical choices. What seems debatable, and a matter of debate even within his work, is how this bleak vision gets mapped onto society. Instead of seeing society as full of social purposes, hence full of significance (in the lesser, subphilosophical sense of the term), Weber tended to divide society into ordinary folk who denied the ultimate emptiness and the scholar or sociologist who, set apart from society, heroically em-bodied that emptiness. Weber's strongest denial of the need for meaning sets the scholar against the theologian. Every theology, Weber declares in "Science as a Vocation," "presupposes that the world must have a *mean-ing*."[12] The scholar ought to be able to live without this theological prem-ise. Here Weber strikes the tone that Ringer describes: "To the person who cannot bear the fate of the times like a man, one must say: may he rather return silently, without the usual publicity build-up of renegades, but simply and plainly. The doors of the old churches are open widely and compassionately for him" (155). To return to religion will entail a "sacrifice of the intellect." Weber does not speak against this sacrifice. He seems to be positing that only a few will have the courage not to offer it, and by his very generosity raising the prestige of the nonsacrificing few still higher. But the passage strikes a general note ("the person" = *Wer*) that allows it to be read as asking everyone, not merely scholars, to bear the fate of meaninglessness manfully.[13]

This democratic hint suggests another, more positive way to under-stand Weber's take on meaninglessness. If meaninglessness is indeed a so-cial and historical phenomenon (as suggested by "the fate of the times"), then there will be more to say about the social history that explains it. Weber lays out this history in a paraphrase of Tolstoy. Tolstoy argued, Weber says, that for "Abraham, or some peasant of the past," death is meaningful because at the end of his days he had received "what life had to offer," because "for him there remained no puzzles he might wish to solve." Today, on the other hand, "there is always a further step ahead of one who stands in the march of progress" (140). For the civilized man, puzzles remain that we do wish to solve. What the civilized man experi-ences is thus "provisional and not definitive, and therefore for him death is a meaningless occurrence. And because death is meaningless, civilized life as such is meaningless" (140). Here meaninglessness seems to be a direct and inevitable result of the belief in progress, which takes the form of puzzles left unsolved.

But Weber's premises lend themselves to a very different conclusion. Recall that enchantment for Weber has not disappeared, but has only been relocated: "the ultimate and most sublime values have now retreated from public life either into the transcendental realm of mystic life or into the brotherliness of direct and personal human relations" (155). The site of the loss is "public life." It is because meaning has supposedly departed from public life that there is no longer "genuine community" (155). But the historical assumptions here seem anything but sure. In order for ultimate values to have retreated from public life, they first had to have resided there. Was public life ever invested with meaning? Under what conditions can "genuine community" be said to have existed? Weber does not ask these questions, but when he puts "puzzles" and "public life" so close together, he invites them. After all, the unsolved puzzles that haunt people like Weber, keeping them from a fulfilled or satisfied death (Weber himself was to die within a year of the publication of his lecture), were *social* puzzles, and puzzles that would arguably have been just as relevant to premodern society as to his own. The suggestion that "genuine community" used to exist omits any consideration of those who were excluded from such community—the slaves and women of ancient Greece, to pick at random from a long series of possible examples. Opinion on the genuineness of community would depend on whose experience was consulted. If you asked landless laborers in the Middle Ages, such community might have seemed less unquestionably genuine.

But the crucial step in the argument is still to come. It's this: if genuine community does not seem to exist in the present, as the disenchantment story suggests, it's in large part because people like Weber *are* asking the marginal and the excluded. In effect, with allowance for the different methodologies of the time, *that's what the new sociology does.* The "puzzles" sociologists work on and that remain unsolved at their death involve questions of what makes public life public, what stops community from being genuine; they are puzzles of social inclusiveness, social cohesion, social justice. It is trying to solve these puzzles, Weber says, that puts one on "the march of progress" and thus leads to meaninglessness. But it makes more sense to conclude that meaninglessness is produced not by progress, but on the contrary by the *failure* of progress: that it is a symptom of the present's failure to integrate social constituencies that the premodern world made no effort to integrate, the present's failure to achieve a level of social justice that the premodern world did even not strive to achieve. In other words, the puzzles are not new. What is new is the attempt to solve them. If progress entails or includes progress toward democracy, toward greater and more equitable social inclusion, then progress toward solving these puzzles is exactly the opposite of what Weber says it is. Progress is not the cause of meaninglessness, but an attempt to cure it. This is just what Weber's puzzle-solving work is doing: not disenchanting a world that would have been enchanted without it, but on the contrary trying its best to supply meaning and community that had never yet existed.

This line of argument offers meaning to Weber's work—in fact, greater meaning than Weber himself is willing to claim for that work. It suggests that, for all his mandarin suspicion of progress and democracy, he himself is working toward something like democratic progress. It suggests that in so doing he is working against disenchantment, and this although he prefers to see disenchantment, most of the time, as unalterable. It suggests, finally, that something can be done to level the supposed "meaning gap" between scholars and nonscholars: that making ordinary life more inclusive and more just will provide ordinary people with some of that "meaning" they are said to hunger for, while it also makes more sense of the scholars' existentially empty labors, reducing their necessity to face meaninglessness in proud solitude.

Do we want to call this reenchantment? I don't see why we would. The hint of delicious, delirious intoxication, while always welcome outside working hours, seems out of place here. This meaningfulness is more prosaic, more everyday, an extension of the meaning to be found in helping a child with her homework, making a good meal, or eating that meal in good company. It suggests that enchantment or reenchantment is the wrong term because disenchantment has never happened. Was there in fact a historical moment when people lost their belief in helping children with their homework or offering good hospitality? It's a short step from these questions to another, which Weber comes closer to posing: did people ever lose their will to join together with others and try to change their shared conditions of life?

His implicit no to this question, implying that disenchantment cannot, then, have happened, is articulated in one of the very passages where Weber announces disenchantment:

> Different gods struggle with one another, now and for all times to come.
> We live as did the ancients when their world was not yet disenchanted of
> its gods and demons, only we live in a different sense. As Hellenic man at
> times sacrificed to Aphrodite and at other times to Apollo, and, above all,
> as everybody sacrificed to the gods of his city, so do we still nowadays,
> only the bearing of man has been disenchanted.... Fate, and certainly not

'science,' holds sway over these gods and their struggles. One can only understand what the godhead is for the one order or for the other, or better, what godhead is in the one or in the other order. With this understanding, however, the matter has reached its limit so far as it can be discussed in a lecture-room and by a professor. (148)

Figuratively at least—but it's a figure Weber won't let go of—the social order that the scholar-teacher steps out of when he enters the classroom is still polytheistic. In this sense, full disenchantment has not occurred, despite the significant change in "bearing" from premodern to modern society. Nor is disenchantment going to happen. With the phrase "now and for all times to come," Weber comes very near to making polytheism the eternal law of human life, thus taking his leave of the disenchantment narrative altogether. The gesture is welcome. It's a relief to imagine being spared the wishfully, willfully blind absurdities of this particular "before" and "after," to imagine that after all there has been no enchantment and there is now no meaninglessness, but rather a persistence of the *kinds* of dilemmas society had suffered through before, however different the dilemmas themselves or the form they take.

Yet perhaps the hinted departure from narrative is also a bit premature. Weber describes "the fundamental fact" of polytheism as follows: "as long as life remains immanent and is interpreted in its own terms, it knows only of an unceasing struggle of these gods with one another. Or, speaking directly, the ultimately possible attitudes toward life are irreconcilable, and hence their struggle can never be brought to a final conclusion" (152). One might be tempted to think of the element of theism in this polytheism as a mere figurative token. But it isn't. It guarantees the uniqueness and irreducibility of each god, hence the irreconcilability of the gods with each other. This is the all-important logic by which polytheism for Weber (like "difference" in its strong versions today) comes to be defined by the necessary frustrating of any "final conclusion." Yet recall that for Weber, inconclusiveness (the "provisional and not definitive" quality of individual lives) is precisely what had produced the meaninglessness and disenchantment associated with "the march of progress." Inconclusiveness thus finds itself on two opposing sides: the side of progress, and the side of polytheism. The two terms look like contraries for Weber; it is polytheism, "an *unceasing* struggle of these gods with one another," that makes progress seem inconceivable. But by associating progress with inconclusiveness rather than with triumphalism, he seems to be drawing them closer together, making it possible to imagine a narrative, which is to say a kind of meaningful progress.

Weber's language here is Nietzschean, and so is the project of social science as Weber envisages it, founded as it is on the purposelessness of the void. It may be that we should simply embrace, with Nietzsche, the antinarrativism of "unceasing struggle" or, in Foucault's version, each regime of domination succeeding the last in an unending display of the implacable ingenuity of power. Still, there is some sense that this turn to Nietzsche may be evading a potential narrative that is offered up by Weber's own vocabulary and that, if acknowledged, could enter into interesting competition with the Nietzschean "now and for all times to come." The figure of polytheism, or the persistence of plural gods, leads directly into a counternarrative of secularization. Weber makes the point himself: "Many old gods ascend from their graves; they are disenchanted and hence take the form of impersonal forces" (149). Here Weber reads secularization not as rupture but as continuity, as a combined preservation and transformation of religious concepts rather than an absolute break with them. This reading does not seem incompatible, though Weber does not say so, with a sort of progress. Impersonal forces are not *the same* as gods. Their authority may, for example, be more vulnerable to nonpriestly intervention, democratic reinterpretation, or flat denial. Disenchantment in this sense produces significant differences, if incomplete ones. The question is how to weigh the transformation against the preservation. These are exactly the materials of linear narrative, which would not be narrative at all if it did not, despite its linearity, loop back to reconnect with its beginnings. A narrative of progress—for example, toward democratic inclusiveness and equality—might well have to forgo conclusive resolution, and even embrace its own (polytheistic) inconclusiveness, and yet still take the form of a narrative with a direction, and still count as meaningful progress. Polytheism and progress together: this would be one way to retell the story of secularization with as little reference as possible to enchantment.

❧

On the first page of the first chapter of *A Secular Age*, the book that won him the Templeton Prize in 2007, Charles Taylor chooses the word "enchanted," despite its evocation of "light and fairies," to describe the world of 1500.[14] He does so, he says, on the grounds that Weber's "disenchantment" has "achieved such wide currency" as a description of the modern world that succeeded it. Hence he can now "use its antonym to describe a crucial feature of the pre-modern condition," namely, "the

world of spirits, demons, and moral forces which our ancestors lived in"
(25–26). The crucial words here are "fairies" and "demons." For the moment at least, Taylor doesn't want fairies *without* demons, good spirits
without evil spirits. This is going to be a full-dress history of the process
of secularization that, like Weber's, will refuse to pretend that the values
of the modernizing progress narrative can simply be reversed. Here, it
seems, the vanished premodern will not stand for all that is good. And,
again like Weber, Taylor will see much of value in our disenchanted modernity. Yet this version of secularization doesn't quite choose continuity
over rupture.[15] Nor can it quite manage to point toward a narrative of
progress, even as an unrealized potential.

Taylor does not want to describe secularism's ascension as what he
calls a "subtraction" story. According to this version of secularization
or disenchantment, religion is said to shrink as science, technology, and
rationality expand. Little by little superstitions are subtracted from the
world, leaving secularism behind to represent all that is not superstition.
For Taylor, secularism is not the widening zone of clarity left as myth and
error are dissipated, but rather the product of shifts in thinking *within*
religion, and in particular within Christianity. In this sense he is clearly
writing in the lineage of Weber's *Protestant Ethic and the Spirit of Capitalism*, which makes precisely this case. One key difference is that Taylor
gives credit for secularization—if credit is the right word—not to Prot-
estantism but to Catholicism. Like Hans Blumenberg, he argues that it
was in the medieval period that the supernatural was first divided off
from the natural, thus preparing the later moment when "an enchanted
world, full of spirits and forces," would become almost incomprehensible to most people.[16] The cause of this early disenchantment was not
protoscientific rationality but rather a more conscious and zealous dedication to God. Taylor's account of the Reformation puts the same irony
on more familiar ground: "So we disenchant the world; we reject the
sacramentals; all the elements of 'magic' in the old religion. They are
not only useless, but blasphemous, because they are arrogating power to
us, and hence 'plucking' it away from 'the glory of God's righteousness.'
This also means that intercession of saints is of no effect. In face of the
world of spirits and powers, this gives us great freedom" (79). Taylor
says that we have disenchanted the world, yet like Weber he also offers
evidence that in an important sense the world is not disenchanted at all.
For Taylor, the modern concept of freedom is Christian (Catholic as well
as Protestant) at its origin, and to an important if unspecified degree it
remains Christian. Modern individualism similarly retains the imprint of

a "Christian, or Christian-Stoic, attempt to remake society" (155). Several hundred pages later, Taylor shows how "the Victorian Christianity of self-discipline created a space for the move to a humanism of duty, will, and altruism" (396). And what is true of humanism is also true of post- or antihumanism. Many of us now look out upon a vast, indifferent universe whose order, such as it is, seems to have nothing to do with the hopes and fears of our self-important little species. But we have arrived here, Taylor proposes, not only because our ancestors abandoned "the immediate encounter with spirits and forces," but because they did so in favor of a "much more powerful sense of God's ordering will" (375). And that sense of order, or one very like it, continues to inform concepts we think of as secular. From a viewpoint that would like to be more rigorously secular, then, there is some question as to whether contact with the indifference of the universe has yet been successfully made, or ever will be.

Writing as a practitioner and defender of Christianity, Taylor cannot be entirely happy to present Christianity as a kind of vanishing mediator, less a form of enchantment in its own right than an agent responsible for leading us all the way from enchantment to disenchantment, the end point that is, or that he fears will be, Christianity's vanishing point. But this is at least one of the stories he tells. When any given religion becomes merely one option among many, he argues, no religion can remain what it was. This is what has happened to all Western religions in the secular age. Taylor calls this secular terminus the "immanent frame"—a phrase that has become the title of a spirited thread of discussion on the Web site of the Social Science Research Council, one of the foremost sites for the "return of religion" on the American intellectual landscape. The immanent frame is "a 'natural,' or 'this-worldly' order which can be understood in its own terms, without reference to the 'supernatural' or 'transcendent,'" and that therefore discourages the choice of supernatural or transcendent explanations. "This is something we all share," Taylor says (594). He does not exempt himself or other believers from it. Here he echoes his earlier position on the inevitability of secularism, a position for which he was strongly criticized by the anthropologist Talal Asad, one of the most influential antisecularists now writing.[17]

The generosity of this concession, so dramatically at odds with where Taylor wants the narrative to go, displays a certain unsettledness at his story's heart. For if he tells us that we are now blocked off from the supernatural and the transcendent, he is no less committed to the notion that try as we might, we have in fact never left the supernatural and the transcendent behind. In other words, the book seems drawn toward two nearly antithetical alternatives. On the one hand, it suggests that secularism's triumph over Christianity, like Christianity's over the pagan spirits and demons, does mark an unquestionable rupture. That triumph, embodied in the immanent frame, now makes religious belief in the old sense very difficult. On this reading, the disenchantment of the world is an unhappy if not quite an irreversible historical fact. On the other hand, *A Secular Age* also presents secularism as a disguised form of Christianity, hiding theological content behind apparently secular concepts. On this reading, the disenchantment of the world was more like a translation. Everything depends on the translation's faithfulness, so to speak— on how much of Christianity was lost in the process and how much was preserved. Taylor acknowledges neither the irresolution produced by the latter reading nor the difference between the first line of interpretation and the second. He simply vacillates between them. His vacillation means that he is more affirmative toward secularism than might be expected from a champion of Christianity. And it means he is more ambivalent than might be expected toward Christianity itself. He sometimes describes secularism as an "achievement" (542), and he sometimes seems to blame Christianity for wiping the old pagan world clean of its wild bunch of demons and spirits.

One of the book's effects is to level the playing field between secularism and religion. This involves both invoking history and flattening it out. Sounding unrepentantly Hegelian, Taylor suggests that the standard of truth does not apply to entities which are subject to history. As a historical extension of the religious project, secularism offers a particular set of answers to the unavoidable question of how to live a good life. Reprising the old communitarian objection to liberalism (unlivable because abstract, empty of value), Taylor assumes that *some* answer *must* be given, even though no answer can ever be firmly grounded or incontestable. Since we're not given enough to go on, he writes, "going one way or the other requires what is often called a 'leap of faith'" (550). All answers to such questions are equally groundless—there is no difference between being "for" something and "believing in" something. Religious belief gets an epistemological pass, therefore, and secularism has to surrender its sense of superiority, becoming the same *kind* of thing as what it defines itself against. This might be thought of as a postmodern case for religion, for it takes as given a general lack of foundations. At the same time, however, we are only a short step away here from the fundamentalist claim that secular humanism is "just another religion" and that Darwin's account of creation deserves no pedagogical preference over the Bible's.

As it happens, Taylor is not friendly to fundamentalism. He has no time for Hell or for the wrath of God. (His God is a God of love, he says, not a "cruel puppet-master" [389]. When the subject of suffering arises, he appears to suffer too, in sympathy, but like his God he is evasively silent on the whys and wherefores.) Taylor himself seems a bit wrathful that the fundamentalists, by accepting a "normal" sense of causation, affirm the materialist concept of the miracle as "a kind of punctual hole blown in the regular order of things from outside" (547). He speaks strongly against churches that identify faith with codes of sexual behavior. What he calls "code-fixation" (704) is one parallel between his counter-Enlightenment narrative and that of Foucault. For Taylor, Foucault's "disciplinary society" stands as a useful description of the secular present—a present that the fundamentalists unwittingly accept too much of when they embrace a "fetishism of rules and norms" (742), especially sexual ones.

The problem is that much of what Taylor dislikes about fundamentalism also applies to Christianity in general. After all, it, too, has moved on from that "enchanted world, full of spirits and forces" about which Taylor does not seek to hide his nostalgia. (The first time the enchanted world comes up, it's described as "the world of spirits, demons, and moral forces" [26]. Later the demons drop out. I can see why.) In one mood, Taylor posits the functional equivalence of the secular and the religious; in another, more driven by his nostalgia, he emphasizes on the contrary a dramatic historical rupture between them—a rupture that is also *within* religion, separating the enchanted world (now demon-less) from Christianity itself. Here he has no doubt that secularism has indeed disenchanted the world and that this is a very bad thing. We have not only lost the spirits, which required some sense of a "higher" reality, but we have also lost our full experience of our bodies. Now we take "a distance from our powerful emotions and our bodily functions" (139).[18] Our religions are no longer really religions. We are barely capable of imagining how much else we may have lost.

Imagination is a sort of magical helper in this story, happened upon when Western man seems most lost in the wasteland of disenchantment, and offering some hope of a happy ending. What if the spirits had not disappeared after all? What if disenchantment were only "an end to porousness in relation to the world of spirits" (137)? The idea that instead of being properly porous, we are now pathologically "buffered" suggests that the world of spirits is still there to be seen; the ability to see the spirits depends solely on shedding our buffers, cleansing our imagination or our epidermis, opening ourselves up body and mind to the higher as well as the lower world. (Sin, as Taylor describes it, is closing oneself off to God; accepting the immanent frame means accepting a closed world. Closure is one of his most frequent and most ideologically loaded diagnoses.) Here Taylor seems to assume a proposition—the real existence of spirits—that reasonable people might doubt and that therefore ought really to be supported by some sort of argument. The only excuse I can imagine for his doing so is that Taylor is also assuming that the widespread *belief* in spirits can be equated with the *existence* of those spirits. In other words, he is giving decisive ontological weight to subjectivity, feelings, imagination.

This is the problem with his characteristic use of "we" and "our." "Our" feelings count as hard evidence that various questionable entities have a real existence. In support of "the idea that nature has something to say to us," Taylor offers those "feelings of renewal" that sometimes come to people in countryside or forest (358). Well, yes, wonder is wonderful, but do people really have to go to the countryside or forest to feel it? And do they feel it there (rather than walking down a crowded city street, for example, or watching someone rap or make sushi) because nature is really talking to them, or because they've been told what to hear and where to hear it? Feelings are always interesting but never reliable. Or again: "I cannot see the 'demand for religion' just disappearing like that" (435). Taylor says this as if once the heartfelt demand for religion has been expressed, the ultimate satisfaction of that demand is somehow obligatory and guaranteed. I feel that God exists; therefore he does. We desire eternal life, and therefore eternal life must be real. Why not conclude on the contrary that because we fear mortality, there is no afterlife? We feel a lot of things. Many of them contradict each other. They can't all be true.

On the one hand, Taylor's respect for certain human feelings seems exaggerated. On the other hand, there are all sorts of feelings that don't get respected nearly enough. There's little or nothing here about secularists finding meaning by helping children with their homework or cooking good meals, about men campaigning to protect doctors from murderous antiabortion activists or Jews campaigning against Israeli settlements on the West Bank. Taylor's portrait of secular modernity is full of stale *Brave New World*–style cliché about Hugh Hefner, brightly lit supermarkets, empty suburbs, and the triumph of the therapeutic. Things get a tad aggressive. Secularists, we are told, are utopians and protofascists by nature. And their lives, of course, are meaningless. Taylor is quite taken with the cliché that life in today's secular world is beset with the malaise of meaninglessness. He repeats it without wondering whether, to the extent that it exists, it might be a result of rising expectations rather than disenchantment—a product of democratic progress to be set against centuries of resignation by the poor to their inevitable social fate. No, there was no malaise back then. Why? Because people knew their places. Nor does Taylor bother to compare this putative malaise with the various sorts of sickness, figurative and literal, that people suffered through in the meaning-saturated medieval parishes that he is fond of evoking for contrast. The demons are not scary enough, and in any case they are casually omitted from ensuing lists of spirits, fairies, moral forces, and so on. There is not enough of the fear that vulnerability to literal rather than figurative enchantment would naturally elicit. There is not nearly enough about the ordinary bonds of work, family, play, and politics, the newly invented intimacies and the technologically mediated attention to distant others, the infinitely varied and surprising forms of love and hope and tenderness that, despite a state of social emergency and the lack of any transcendental foundation, provide most of us, most of the time, with enough meaning to go on.

In short, Taylor is telling the disenchantment story again, and telling it with a vengeance. This story makes life without religion/enchantment seem so poor and depleted, so toxic and utterly unlivable that one is almost ready to embrace Taylor's less nostalgic, more postmodern alternative: that the world is ruled by feeling and imagination. But that, too, would be a mistake.

The immanent frame (that is, secularism) is what Taylor calls a social imaginary. Taylor understands the social imaginary, the subject of one of his earlier books, as a set of normative notions and images that go deeper than mere "intellectual schemes" and that determine the sorts of expectations that are and aren't possible. It does not specify beliefs, but sets the conditions of belief. Crucially, it determines how much harder it is to sustain your faith once your faith comes to be seen as merely one option among others. But how determining you think a social imaginary is will depend on whether you accent the social or the imaginary. If you accent the social, you get the suggestion that for better or worse secularism is not going away—the tiny, isolated medieval parishes that once sustained belief are not coming back. If on the other hand you accent the imaginary, you can be optimistic again about belief, but you are suddenly in doubt as to whether this age is indeed secular at all. To say that everything is imagination is to say that everything is belief. If that's what Taylor is arguing, then the disenchantment story collapses. We believe now, he would be saying, and we have always believed. Belief is not impossible; it's all there's ever been. Secularism is merely the arrogant illusion that there can be anything other or higher than belief.

Taylor mobilizes Nietzsche, the "death of God" man himself, in this attempt to cut secularism down to size. Perverse as the strategy may seem, it makes a certain sense. When Nietzsche declared that God was dead, he went on to add, at the risk of paradox, that this meant God had in fact gone into hiding and now had to be smoked out of various secular terms, from morals and nature and history to man and even grammar. The large question that Nietzscheans like Foucault have never been very good at facing is whether these God-terms are or are not God-equivalents. For if the secularization of theological concepts results in nothing but more theology, if God-terms are functional point-for-point equivalents for God, then God is effectively indestructible. In that case, everything would indeed be religion, or faith, or imagination, or belief. Taylor flirts with this idea, and sometimes seems ready to espouse it. "Perhaps there is only the choice between good and bad religion" (708).

No one can be entirely happy with the desperately ahistorical postulate that "everything is religion," however gratified one may be to observe that it contradicts Taylor's history of disenchantment. Secular readers will want to insist on maintaining a few distinctions: for example, Darwin's theory of natural selection does not belong to the same generic or epistemological category as does the biblical story of the Creation. Literary critics like myself will perhaps want to rethink their own self-congratulatory versions of the "everything is X" equation. What I hope everyone will want is a history that presents the transition from God to God-terms—that is, a history of secularization—as real and significant, even if God-terms always invite further suspicion and further secularization. After all, one need not be satisfied with a quasi-theological use of "nature" or "history" or "human rights" in order to feel that these terms mark a significant improvement over the vocabulary of religion as such. One need not idealize what counts today as rational argument in order to judge rational argument an improvement over appeals to the authority of sacred scripture or to the will of God. Such an improvement story could subtract the nostalgia from Taylor's tale of disenchantment without thereby becoming another story of secularism-as-subtraction—without refusing, that is, Taylor's valuable point that secularism, too, is constituted and limited by history. In short, I think we need a story of secularism as improvement over religious belief, a story that would be both inconclusive and inspiring. At the risk of seeming to reinvent a very creaky wheel, I would call this a narrative of progress.[19]

❧

In *Darwin Loves You: Natural Selection and the Re-enchantment of the World*, George Levine agrees "that Weber is right to recognize the power of 'rationalization' to demoralize."[20] I, too, agree. The question is not whether it would demoralize us if it were the case; the question is whether it *is* the case—in other words, how well the concept of rationalization does in the competition of many rival concepts to decide which best describe the world. I don't think it comes close to winning. The paradigmatic institution of rationalization is bureaucracy. In the midst of a financial and economic crisis that is extraordinarily severe but historically far from unique, hearing news every day about the sufferings it has caused, we are reminded very forcibly of two things that we ought to have known before: that capitalism has more to do with the shape of our world than bureaucracy does, and that although capitalism also uses numbers, it is not a force for rationalization, but a force for chaos. Faced with such wild, irresponsible, number-based imaginings as the derivative, bureaucratic regulation is very emphatically not a useful name for what's wrong with our world. It is something that we needed much, much more of. In short, disenchantment is the wrong diagnosis. And reenchantment is the wrong remedy.

The call for reenchantment cannot help doing precisely the opposite of what it wants to do. It seems to reject the disenchantment story, but it accepts too much of that story, for ordinary life must first be hollowed out and impoverished in order for reenchantment to be granted the contract to fill it up and enrich it again. Wonder and surprise produce the problem they say they want to solve: in Levine's words, "a consistent undervaluing of contemporary experience" (37). The crucial move here is to present ordinary life as "routine," which allows wonder and surprise (that is, that which is not routine) to sound like solutions to a problem. This is a terrible failure of imagination. Since when was life in history, fully experienced in all its dialectical twists and turns, ever a matter of predictable routine? If all biology is miraculous, properly considered, then surely the same is true for human history and indeed for everyday social existence. Is it not, Slavoj Žižek asks in *On Belief*, "that ALL religion, ALL experience of the sacred, involves—or, rather, simply IS—an 'unplugging' from the daily routine? Is this 'unplugging' not simply the name for the basic ECSTATIC experience of entering the domain in which everyday rules are suspended, the domain of the sacred TRANSGRESSION?"[21] Well, no. If this capitalized ecstatic transgression is suddenly on the agenda, it's only because people are willing to buy into a depleted, monolithic, contradiction-less sense of "daily routine," a sense that in the normal course of things, the "rules" unfortunately do work. They don't. Anarchy is not what we want, but what we already have. Wall Street traders have spent decades happily transgressing, and their unplugged ecstasies have resulted in foreclosures and unemployment on a massive scale. Capitalism is not rationalization. It does not disenchant the world. What it does is different, and much worse.

As Jane Bennett rightly notes, there is a form of enchantment even in commodity fetishism. But consider how much enchantment or poetry or meaning is also to be found in everyday practices of de-fetishization. When you look at the shirt on your back or the coffee in your cup and, for once, see through the object to the labor of the people far away who produced it, to the lives they are obliged to live, to the invisible but real links between their lives and yours, you may not find enchantment in their lives, but you cannot conclude that the world is dull. I cannot see the merit in making yourself blind to what's in your path in order to ensure that you will be perpetually surprised. To decide that everyday life is rationalized, bureaucratized, or routinized is to kill it in order to get a pat on the back for rescuing it from the dead.

I cannot speak to the question of how many of the needs traditionally thought to be inseparable from religion might be satisfied by a move from the routine/wonder narrative (itself now something of a routine) to a more dialectical alternative in which the surprises are to be expected and secularization retains the surprising power of counting as progress. I assume that some of what are called "needs" ought to be thought of as wishes, wishes for things we have never had and cannot have, like a good answer to the "meaning of life" question or a basis in nature for human ethics or a satisfactory consolation for the death of a child. To appreciate what Levine calls the "deep value of everything" (259), including the virus that killed your daughter, seems to me a religious stance rather than an ethical one. It devalues the word "value" itself. There is no adequate recompense for suffering and injustice that have already happened. But (why be afraid to repeat the banality?) there is meaning to be found in trying to diminish the amount of both that will happen in the future.

It would not be hard, I think, to connect secularization to the theme of democratic inclusiveness and autonomy that I discussed above. If there has been a new tolerance for the "return of religion," perhaps there is also increasing room for the recently unfashionable metanarrative of emancipation. Imagining the world without religion, or with less of it, would mean getting out from under an enormous weight of onerous authority. This emancipation is not easy to imagine, since secular concepts do contain so much religious baggage (hence the inconclusiveness of secularization) and so few places are secular even in the most limited sense. In the United States, which is not one of those places, imagining further emancipation would mean, for starters, imagining presidential campaigns without the obligatory display of some sort of faith. It would mean imagining that people might little by little be able to give up saying after a public or private disaster that their suffering is "the will of God," and this although there is no other consolation that can effectively take that formula's place. It would mean people taking their fate into their own hands without any guarantee that things would therefore work out better than they have.

Even without guarantees, a secularization-as-progress narrative may sound arrogant or complacent. But if it is, it is less so than a reenchantment of the world. A modest uncertainty about the limits of human knowledge and action seems to me to fit better on the secular than on the religious side of the aisle. To be secular is to know that not all mysteries can be transformed into puzzles, let alone solved. It means that our words should not ask for what they can't have. If it makes sense to speak, as Levine does, of "the sacredness of a world without a telos" (261), surely it makes more sense to speak of the profanity of such a world. Assuming we want as far as possible for our words to have meaning, which is one of the more feasible ways to ensure that our lives, too, are as meaningful as possible, it might be better to give up on the sacred, and on enchantment in all its variants. There are other and better ways to say nice things about the world.

# 4

## ❧

# Shock Therapy, Dramatization, and Practical Wisdom

### *William E. Connolly*

Understanding, hypothetical or instrumental reason, speculative reason, practical reason, aesthetic judgment, teleological reason. The Kantian list is familiar. Its divisions and priorities are culturally entrenched, even among many who do not confess Kantianism. Each office is supported and sustained by the others and by arguments that specify it once the divisions are set. What if you seek to break that frame in order to create space for an alternative? Why would one want to do that? How would you proceed? The "why" question is difficult to answer in advance, since it consists of suspicions and hopes to be redeemed later through positive formulations. Suffice it to say that I suspect that Kantian and neo-Kantian complexes function to inhibit creative, exploratory experiments in thought and practice, to squeeze explanatory projects into too narrow a compass, to diminish our awareness of the diverse ways the nonhuman world enters into human life and affects our attachment to existence, to define instrumental reason too sharply, and to obscure a needed dimension of ethical life. All this makes it more difficult than otherwise to pull presumptive care for the diversity of life and the fecundity of the earth to the forefront of practice. It also distracts attention from our participation in a larger world of becoming that is in itself replete with differential powers of creativity. I also admire the Kantian system, however, because of how its systematic character challenges other theories to clarify and justify themselves, and because of a care for the world that circulates through it.

My strategy is to move on two fronts. On one front you compress the arguments in support of Kantian entrenchments, doing so to identify "flashpoints" at which key existential investments enter the complex, sometimes unconsciously and sometimes when that juncture is treated by the theorist as an undeniable starting point of everyday experience. On the other you engage in shock therapy to help to *dramatize* those same flashpoints differently, doing so at first by sketching another cultural background from which experience and experiment proceed. The idea is to allow those microshocks to open a door to exploratory thinking. The focus in this essay is on practical reason, the mode that Kant takes to govern morality and to provide the lynchpin of reason in the larger sense. We start with a dose of shock therapy.

## Theogonical Wisdom

Daughters of Zeus, I greet you: add passion to my song, and tell of the sacred race of gods who are forever, descended from Earth and starry Sky, from dark Night and from salty Sea. . . . Tell how in the beginning the gods and the earth came into being, as the rivers, the limitless sea with its raging surges, the shining stars, and broad sky above. . . .

The great Cronus, the cunning trickster, took courage and answered his good mother . . . : "Mother, I am willing to undertake . . . your plan." . . . Then from his ambush his son reached out with his left hand and with his right took the huge sickle . . . and quickly sheared the organs from his own father and threw them away, backward over his shoulder.

Rhea submitted to the embraces of Cronus and bore him children with a glorious destiny: Hestia, Demeter . . . , Hera, Hades . . . and Zeus the lord of wisdom . . . , whose thunder makes the earth tremble.

She (Earth) took him [Zeus, the youngest son of Cronus] and hid him in an inaccessible cave, deep in the bowels of the holy earth. Then she wrapped a huge stone in baby blankets and handed it to the royal son of Sky (Cronus) who was king of the gods. He took the stone and swallowed it into his belly. He did not know that a stone had replaced his son.

When the Olympian gods had brought their struggle to an end and had vindicated their rights against the Titans, Mother Earth advised them to invite Zeus . . . to be king and lord over the gods. . . . Zeus' first consort was Metis (Wisdom). . . . But when she was about to give birth to bright eyed Athena, he deceived her with specious work . . . and trapped her and kept her in his belly . . . so that the kingship would not pass from Zeus to another of the gods.

Lastly Zeus took Hera (his sister) as his wife to bear him children. . . . Likewise Semele, Cadmus' daughter, lay with him in love and became the mother of a son with a glorious destiny—Dionysus the giver of joy. She was mortal when she bore her immortal son; now they are both immortal.[1]

What a world! Several points differentiate it from mythic/spiritual determinations that infiltrate Kantian and neo-Kantian philosophies. First, the gods may live forever, but they defeat each other periodically and they fornicate with humans rather often. Second, the idea of cause in that world is not reducible to efficient causality. Sex, love, sensuality, and deceit provide better metaphors from which to think causally than the idea of humanly designed mechanisms from which the idea of efficient cause is drawn. Absorption, swallowing, intermingling, digestion, and strengthening are all modes through which defeat, transformation, and transfiguration occur. Third, the gods are multiple and not entirely subsumed under a single principle or historical trajectory. Fourth, the modes of interplay between gods and humans are also multiple, with some humans becoming gods. And—as the tragic playwrights who later work upon the *Theogony* emphasize—such gods or forces both enter into human passions and operate upon events from the outside to support a sweet victory or tragic result. Fifth, the late introduction of Dionysus into the mix both points to an element of wildness in nature/culture imbrications and appreciates the sweetness of life in that very world. Dionysus, himself the result of an illicit crossing, speaks to a Greek readiness to join together the element of wildness and joy in the human condition.

When you combine these five points, you can see that the cosmos to which Hesiod is attached is neither deeply providential nor receptive to consummate human mastery, even if it does become a bit more tidy with the Olympians. It thus did not take all that much for Sophocles to transfigure this myth later into a tragic vision, as Athens confronted its own conflicts between old and new gods. Conflicts, surprising turns, and unexpected events periodically punctuate the regularities of civil life, steady tradition, and ethical precept, creating new issues for decision and judgment. The result is not "chaos," as some devotees of the straitjacket image of order love to say whenever they encounter a vision identifying a whiff of chanciness in life: *rather, the world consists of durable periods of relative order punctuated by periods of acceleration in which disruptions occur, owing in part to forces beyond human agency.*

I said "gods or forces." As Jean-Pierre Vernant has shown, Ionian philosophers such as Anaximander and Democritus did not have to work that hard to translate these intermingling and contending gods into multiple forces. Hesiod had already opened that door with his notions of primordial Earth, Sky, Cronus, and so on. "The fundamental concepts that the construction of Ionian philosophy is based on—the separation from a primordial unity, the constant struggle and union of opposites, and an eternal cycle of change—reveal the ground of mythical thought in which Ionian cosmology is rooted. The philosophers did not have to invent a system to explain the world; they found one ready-made."[2] These latter ideas also surface as minor themes in Sophocles, as when Jocasta explains to Oedipus the active role of chance in life and the cosmos, before he, the chorus, and (most of?) the audience later interpret their tragic fate to be anchored in the hostility of the gods.

What would practical reason, as rules and dispositions appropriate to moral life, look like if our world had evolved directly from this one rather than detouring through two thousand years of Christianity? What about instrumental reason? There is no reliable answer to these questions. So the counterfactual will seem badly posed. Nonetheless, I pursue it. A genealogy of the present pursues such counterfactuals to expose and disturb unconscious presumptions, feelings, and insistences that infuse argument and judgment now.

With that caveat, we can suggest, first, that the Kantian idea of the understanding would be pushed into crisis, since the idea of efficient cause is too simple and schematic to grasp the interplay, infusions, and transfigurations that constitute relations among gods, humans, and nature in the *Theogony*. As Michel Serres has shown in his attempt to "modernize" early Greek notions of science, the ideas of fluids and flows were central to them, while those of mechanics and stasis have been given more priority by moderns, at least until recently.[3]

Second, the idea of practical reason would now devolve into a quest to inculcate wisdom into everyday affairs in a world not intrinsically designed for either human benefit or human mastery. Cultivation of the right kind of character also becomes a matter of prime importance, partly because there is no characterless moral subject in this world issuing universal moral laws, and partly because the larger world is not postulated to be predisposed to humanity in the last instance. Here I concur with Bernard Williams, in his review of similarities between the classic Greek tragic vision and operational ideas of modern life partly obscured by dominant philosophical accounts.[4] Negotiation of the right cultural ethos, for the same reasons, would attain similar importance for the quality of public life.

Instrumental reason would be touched too. In the Kantian tradition it is demarcated by contrast to the purity of practical reason and aesthetic judgment. It is calculative action guided by a set of "sensuous" interests not limited by supersensible moral considerations. But if character is critical to practical wisdom in the world of Hesiod, it also makes a difference to the definition and pursuit of self-interest as you pursue power, income, sexual liaisons, and self-security in that world. The infusion of practical reason with elements of sensuous character and close attention to the fragility of things in a nonprovidential world would thus combine to suggest a reformulation of Kantian, Habermasian, and neoliberal notions of *instrumental* reason as well as of practical reason. When you calculate your interest and the means to it, elements of character and tradition invested in you slide into the calculus, carrying it along some tracks as opposed to others. Some of the elements of character that now enter practical reason also infuse instrumental reason, blurring the sharp boundaries between them set up in Kantian philosophy.

To bump these considerations into a post-Kantian register, you could say that the memory-saturated "somatic markers," which the neuroscientist Antonio Damasio identifies as nonconscious modes of memory that prime and narrow the range of options for calculation before conscious reflection, play a constitutive role inside "instrumental reason."[5] Suppose you belong to a political science department. A colleague first campaigns militantly against a candidate and then, upon losing that round, campaigns just as militantly to put the candidate up for tenure immediately, against the advice of her supporters. Kill the candidacy this way or that. Is that an instance of "pure" instrumental reason? Not really. Even if we ignore the long-term effect of such transparency, a disposition to ruthlessness has entered the calculus, turning the judgment differently than would be the case if another disposition had infused it before decision. If character is ubiquitous to ethical judgment, there is no such thing as pure instrumental reason either. For there is no judgment without character, and the contours of this indispensable component vary from person to person (and place to place).

But what about aesthetic judgment? My engagement with Kant will focus on *The Critique of Practical Reason* and *Religion within the Limits of Reason Alone* because it is there that the themes most resistant to the onto-political perspective developed here are adumbrated. But in *A Critique of Judgment* Kant does ventilate his system to some degree. There he speaks of aesthetic judgment as preconceptual *and* governed by an implicit concord of the faculties. He focuses a bit more on the *receptivity* of the subject to the world than on its pure *constitutive* powers, and he even hints that the subject may be formed to some extent as well as given. Authors such as Alfred North Whitehead and Gilles Deleuze have worked on Kant at these points, seeking to bathe the earlier texts in an expanded version of the later themes. Here is the way one recent author responds to those attempts:

> I have been dwelling on Whitehead's self-proclaimed inversion of Kant because I want to suggest that Kant himself already performs something like this inversion or self-correction in the Third Critique. For there Kant proposes a subject that neither comprehends nor legislates, but only feels and responds ... ; this subject is itself informed by the world outside, a world that (in the words of Wallace Stevens) "fills the being before the mind can think."[6]

Such a project is full of promise, though I think it must work upon Kant at least as much as it draws sustenance from him. The adherents of this project tend to compare the adjusted Kant to the earlier Kant, while I seek to compare the early and adjusted Kant both to themes in the *Theogony* and to the maxims of practical reason advanced in the final section of this paper. Those differences in the terms of comparison make all the difference.

What about Hesiod and time? My sense is that there is a pregnant connection between the idea of time as becoming accepted by me in this paper and the orientations to time in Hesiod and Sophocles, even though they definitely do not coincide. Both perspectives play up the possibility of sudden shifts and turns in our experience of time. Each perspective appreciates the character of tragic possibility in human affairs in a cosmos not highly predisposed to the interests of humans. Here is the way Jacqueline de Romilly makes these points with respect to Sophocles:

> Sophocles, of course, knew about divine justice, and about suffering caused by ancient fault. Yet he seldom insisted on the idea. He does not say that the event which comes and destroys man arises from a just or unjust power: he says that it was God's will. And the consequence is that the long delays in divine justice are less dwelt upon than the sudden intrusion of God's will in human life. Even when punishment is mentioned, we find, instead of an impending threat, quickness and contrast.[7]

But I am running ahead of myself. I have not addressed the *arguments* Kant presents to sustain the unity of reason and the complex relations between its offices. Nor have I noted how a neo-Kantian such as Jürgen Habermas, after taking a linguistic turn, modifies the letter of Kantian reason while preserving much of its spirit. It is time, then, to let Kant speak. Is it also timely to probe what embodied cultural predispositions might speak *through* Kant?

## Pure Practical Reason

*God*: Who are we mere men to presume to set limits to his knowledge, by saying that if temporal things and events are not repeated in periodic cycles, God cannot foreknow all things which he makes. . . . In fact his wisdom is multiple in its simplicity, and multiform in uniformity. It comprehends all incomprehensible things with such incomprehensible comprehension that if he wished always to create new things of every possible kind, each unlike its predecessor, none of them could be for him undesigned and unforeseen. . . . God's wisdom would contain each and all of them in his eternal prescience.[8]

*The divided will after Adam's fall*: It does not will in its entirety: for this reason it does not give this command in its entirety. For it commands a thing only in so far as it wills it. . . . But the complete will does not give the command.[9]

*Divine Grace*: Open his eyes then by exhorting him and praying for the salvation he ought to have in Christ, so that he may confess the grace of God the saints are proved to have confessed . . . ; for these things would not have been commanded . . . , nor would they have been asked for, unless to the end that the weakness of our will should have the help of Him who commanded them.[10]

*Eternal life*: This faith maintains and it must be believed: neither the soul nor the human body may suffer complete annihilation, but the impious shall rise again into everlasting punishment, and the just into life everlasting.[11]

What a world! According to Augustinian Christianity, the will simultaneously separates human action from the simple effect of God's creation, protects humans from material determination, explains how the painful rift tragedians found in being itself is actually lodged in a human will divided against itself after Adam's rebellion, and shows how human beings themselves are responsible for the rift in being. Free will protects the omnipotent Creator from responsibility for evil and thereby creates a divinity powerful and benevolent enough to fulfill the Christian hope for eternal salvation. After Adam's first disobedient (free) act, humans can will bad things by themselves, but divine grace is always required if they are to will anything good. The will is thus profoundly entangled in aporias unless divine grace is available to it. We nontheists call that the element of heteronomy within the will. This need for grace, amid profound existential uncertainty about it, flows from faith in divine omnipotence, providence, and an eternal salvation.

Each of these articles of faith reemerges, albeit in modified form, in Kantian reason: the revisions are funneled through it as necessary postulates, hopes, and "as if" assumptions of reason, with each receiving a distinct standing. The *revisions* are needed in part because Kant accepts a Newtonian account of nature unavailable to Augustine, and in part because Kant, drawing upon advances in hermeneutical research, concludes that the sacred texts upon which Augustine had rested his case are filled with problems of translation and issues of "redaction" that complicate their authority. The *similarities* reflect a continuity—though not an identity— in the concepts of morality and faith that Kant now seeks to bring into line with Newtonianism.

What about the postulates, and the like, of practical reason? Do these operations of faith, conceptualization, edict, and hope express cultural predispositions that are folded back into practical reason? Or, as Kant contends, are such iterations secreted only *after* the baseline of practical reason has been set in indubitable experience, that is, in features of human experience itself undeniable by anyone once delineated? Is the unity of reason a culturally circuitous affair? Or the effect of strict argument anchored in indubitable experience?

Comparative attention to the Hesiodic and Augustinian traditions helps to identify *flashpoints* in the Kantian argument. By a flashpoint I mean, in some cases, a fugitive experience in need of selective dramatization to be rendered actual in this or that way, and, in others, a mundane experience taken by a philosopher to set a necessary or apodictic starting point for argument. Each flashpoint provides a gateway for Kantian arguments of necessity in the various offices of reason. There are several. There is the everyday experience of time as succession that helps to prepare the ground for Kant's deduction of the categories of the understanding. There is the "apodictic" recognition that morality takes the form of law, setting the baseline for the postulates of freedom, God, and salvation, and for looser secretions about providence, grace, progress, cosmopolitanism, and an ethical commonwealth. There is the "nonsensuous feeling" that enables respect for moral duty to be *felt* while insulating that feeling from the wayward *sensuous domain*. Without that insulation the purity of morality would be contaminated. There is also the radical distinction, drawn late in the day, between nonorganic nature susceptible to categories of the understanding and organisms that exceed those categories, escape efficient causality, and require the postulate of a higher intelligence pulling the human world toward its highest end. There is also the spontaneous accord of the faculties below conceptual articulation that enables the experience of beauty and makes it possible to expect that experience to be communicated and universalizable. I focus on the recognition of morality as law.

That morality takes the form of law, rather than, say, expressing a preliminary attachment to the earth and the diversity of life, is not something to be proven. It can, in the first instance, only be experienced. As Kant puts it:

> Moreover, the moral law is given as an apodictically certain fact, as it were, of pure reason. . . . Thus the objective reality of the moral law can be proved through no deduction, through no exertion of the theoretical, speculative or empirically supported reason. . . . Nevertheless, it is firmly established of itself.[12]

This fugitive, indubitable experience of morality as law sets a base point from which the deductions of freedom, God, and salvation occur. The freedom Kant deduces is that of a free will that escapes determination by sensuous means. The will can consent to abide by the moral law when it conflicts with sensuous desire, or consent to those inclinations. Under repeated duress of the latter pressure, however, the internal structure of the will itself eventually moves close to the characterization of the will divided against itself given earlier by Augustine. In the cases of some human beings it closely tracks that structure. So, by the time we reach *Religion within the Limits of Reason Alone* we find that

> there is in man a natural propensity to evil; and since this very propensity must be sought in a will which is free . . . , it is morally evil. This evil is radical because it corrupts the ground of all maxims; it is moreover, as a natural tendency, inextirpable by human powers, since extirpation could occur only through good maxims, and cannot take place when the ultimate ground of all maxims is postulated as corrupt; yet at the same time it must be possible to overcome it, since it is found in man, a being whose actions are free.[13]

Not only does Kant recapitulate the Augustinian theme of the will divided against itself; he is thereby compelled to reintroduce grace into his existential equations, if more cautiously than Augustine had. Grace is no longer a preliminary faith of Christianity as such; it is not exactly a postulate of practical reason either. It becomes an indispensable "hope" that exceeds the limits of reason "itself" but is nonetheless important to protect the integrity of practical reason. It is a necessary supplement to the unity of reason secreted but not deduced from reason as a postulate. The audacity of hope.

To secure the autonomy of agents from material determination, we must not only accept a philosophy of the will that breaks the logic of causality operative elsewhere in nature; we must also hope that grace will lift us above the *internal division of the will against itself* if and when we are unable to do so alone. Otherwise what morality requires may not be possible. And it is necessary to the efficacy of morality itself, Kant says, that it be possible to promote progressively what morality requires.

These postulates and hopes also cross into collective life. The Kantian projection of cosmopolitan progress toward an ethical commonwealth is not grounded primarily in empirical evidence of historical progress. These are assumptions we must project to protect and secure a pure morality that is ubiquitous and ineliminable. So when Kant affirms cosmopolitanism and universal progress, he is not primarily saying that the evidence of history supports these developments. He is stating an implication of his moral philosophy: drawing out a collective implication of the apodictic idea of morality as law and joining it to other postulates and projections already in place to secure that recognition:

> I will thus permit myself to assume that since the human race's natural end is to make a steady cultural progress, its moral end is to be conceived as progressing toward the better. And this progress may be occasionally interrupted, but it will never be broken off. It is not necessary for me to prove this assumption. . . . For I rest my case on my innate duty . . .—the duty so to affect posterity that it will become continually better (something that must be assumed to be possible).[14]

"I will thus permit myself to assume"; "my innate duty." This means that it is necessary to make such an assumption about the progress of civilization to secure Kantian morality more than that the empirical historical record supports the assumption, even though Kant does mine the empirical record for supplemental support. (Some theories of Kantian cosmopolitanism seem to slide over the essential point that it follows from his initial sense of morality even more than it reflects a historical account.)

Accordingly, the hope for individual grace now acquires a collective face too. The assumption of collective historical progress is necessary to redeem the idea of moral universality. And that progress, even when an ethical commonwealth has been established to a considerable degree, exceeds the ability of the ethical commonwealth to foresee the long-term effect of collective actions. So it is necessary to project an element of divine wisdom and grace into the trajectory of collective history. Kant says we *must* believe, as a *collective* upshot of the ineliminable idea of morality as law, that "the love (assured to us through reason) of God toward man, so far as man does endeavor with all his strength to do the will of God, will make good in an upright disposition the deficiency of the deed, whatever the deficiency may be."[15]

The list of postulates, hopes, and upshots continues, all bearing family resemblances to the faith Augustine asserts more directly and links to divinely inspired scripture. We may have reviewed enough of them, however, to see how the logic of practical reason and cosmopolitan progress grows out of a seed in the subject that is said to be apodictic. It may also be pertinent to see how neo-Kantianism, which has often proceeded without reference to such postulates and hopes, now increasingly acknowledges its dependence upon them. For instance, Habermas, who earlier translated Kantian reason into a philosophy of language with unavoidable counterfactual assumptions, now inserts into it an "as if" supplement of Christian evangelism reminiscent of Kant. He does so to insulate human motivation from biological explanation and intervention understood by him in reductive terms. He says:

> Because he is both in one, God the Creator and God the Redeemer, this creator does not need, in his actions, to abide by the laws of nature like a technician. . . . From the very beginning, the voice of God calling into life communicates within a morally sensitive universe. . . . Now, one need not believe in theological premises in order to understand what follows from this, namely, that an entirely different kind of dependence, perceived as a causal one, becomes involved if the difference assumed as inherent in the concept of creation were to disappear.[16]

It seems to me that Habermas, in this late work, moves a step closer to Kant himself, both articulating a notion of scientific understanding too close to his and then adopting an "as if" postulate to protect human beings from reductive explanations and modes of biomanipulation.[17] But complexity theory in biology and neuroscience has gone well beyond the reductive account that requires such an "as if" logic. Some are even prepared to say that we participate in a larger world of real creativity that also exceeds the human estate.

What is an effective way to contest, not the *possibility* of the Kantian logic but its *necessity*? One way is to bring alternative theories and philosophies of causality and time directly to bear on Kantian philosophy. Another, the way pursued here, is to compare the theocosmological background of the *Theogony* to that installed in the Euro-American culture in which Kant participated. A related way is to recall how confessions, devotional practices, church rituals, juridical assumptions, school repetitions, parental inductions, TV dramas, and institutional modes of responsibility and punishment both become infused into such dispositions, however imperfectly, and flow into the higher registers of thinking and judgment. Pursuing those two strategies together, I suggest, discloses deeper sources of all those "as it were" statements Kant makes at pivotal moments ("given as an apodictically certain fact, as it were, of pure reason"). People in Christendom already, as it were, experience the call to be ethical through the rubric of intrinsic law; they already, as it were, demarcate a sharp difference between nonhuman events and human action through the respective concepts of efficient cause and free will; they already, as it were, experience competing dispositions to action as if they are conflicts of the will; they often already, as it were, hope for grace when the divisions of the will are sharp or failures of collective action are severe; they often project, as it were, the assumption of historical progress into life on the pain of otherwise falling into tragic despair. Other supports, earthy affections, character development, existential experiments, and calls to courage are cut off at the pass by such culturally mediated modes of expression.

Keeping the *Theogony* in mind, let's return to that pivotal moment of apodictic recognition upon which so much hangs in Kant's philosophy. Kant himself was inducted into a broadly Augustinian culture, following the pietist tradition of his youth. And young people in Europe, according to him, need to be inducted into it too, to sharpen the fugitive awareness already there that morality takes the form of law, and to prepare the sensuous dispositions to become receptive to the demands of suprasensible reason. Here are a few of his formulations about the induction process:

> Certainly it cannot be denied that in order to bring either an as yet uneducated or a degraded mind into the path of the morally good, some preparatory guidance is needed to attract it by a view to its own advantage or to

frighten it by fear of harm. As soon as . . . these leading strings have had
some effect, the pure moral motive must be brought to mind.

. . . in teaching a man to feel his own worth it gives his mind a power,
unexpected even by himself, to pull himself loose from all sensuous attachments (so far as they would fain dominate him).

. . . the pure thought of virtue, when properly commended to the human
heart, is the strongest drive to the good and indeed the only one when it
is a question of continuous and meticulous obedience to moral maxims.[18]

Kant seeks to induct young people into a cultural mode of re-cognition,
as it were. And, as the pages that follow show, casuistry, exemplars, and
other tactics of induction are commended to sharpen (or dramatize?) the
experience of morality as law and to render the sensuous dispositions
receptive to its dictates. That is, to build the material character appropriate to Kantian morality. It seems, then, that Kantian practical reason
does not simply start from indubitable experience. Rather, in a way that
echoes softly the more receptive image of the subject elaborated in the
Third Critique, it *dramatizes* in a particular way a set of fugitive experiences already installed to some degree in the culture in which he par-
ticipates. It then commends additional tactics and disciplines to fix those
dramatizations securely in the soft tissues of life. But intercoded background beliefs, dispositions, assumptions, and hopes differ significantly
from those installed in the Greek world of Hesiod. Plato's attempts to
dramatize a few Kantian precursors, for instance, required more radical
mythic shock therapy than the stage-fright gimmicks Kant commends.

My way of putting the point is to say that Kant does not amplify an
apodictic human recognition already there; *he dramatizes a culturally embodied seed that could be solicited, amplified, and dramatized in different
ways.* He must obscure the role of dramatization in his account to protect
the aura of necessity surrounding practical reason. The philosopher of
reason is a culturist under the skin who was, as it were, predisposed to
the postulates later deduced from his system. The cultural starting points
from which his arguments proceed are profoundly contestable.

I do not seek to return to the Greek world, but to use it to help rethink this one. I am not content either with the type of skepticism that
is sometimes projected when the tightness of Kantian argument has been
contested. Since there is never a vacuum on the visceral register of subjectivity and intersubjectivity, we are all always already poised to move
in some direction or other. My agenda, then, is to *dramatize* another possibility here and now. To both dramatize that possibility and argue on

its behalf. There are no airtight arguments in this domain. Persuasion, sensitivity to the time in which we live, and attraction or inspiration are all relevant at these bifurcation points.

## Some Maxims of Practical Wisdom

One thing Hesiod, Sophocles, and Kant shared was a care for the way of the larger world as such, as each grasped and engaged it in his way. It is expressed in the style of each, and, as already suggested, in moments of receptivity in each. By care for this world I mean attention to the larger course of things that marks the era in which you live, infused with positive affect toward the most fundamental terms of existence as you grasp those terms. This care for being can be joined to political militance, if and when events threaten the integrity of that which you care for the most. But that militance will also be affected by the sensibility infusing it. That is a lesson I retain from all three, even as I contend that under the global circumstances we face today, an enlarged minority of people must embrace a vision of the world that is neither providentially ordered, teeming with gods, governed by universal moral laws, susceptible to consummate mastery, nor lodged on a secure trajectory of linear progress. A world of becoming set of several tiers of chrono-time, in which the human estate is involved with several other temporal force fields, any of which may impinge upon it in a new way from time to time. The practical wisdom I pursue emerges from these background understandings; it seeks to affirm existentially a world of becoming in which we are minor participants; and it seeks to enter into productive relations with others whose assumptions and maxims diverge in this way or that from those composed here while also expressing a more profound attachment to what Gilles Deleuze calls "this world." To the extent a lived philosophy is infused with existential resentment, it fosters a spirit of punitiveness toward diversity and a refusal to give a degree of priority to the future over the present. It can be highly intelligent, but it does not qualify as a variant of wisdom in my book.

The relations among the elements of practical wisdom embraced here do not assume the standing of entailments rendered necessary by an ineradicable point of departure. The complex, rather, takes the shape of a problematic. The elements secrete and support each other as moving parts of a rickety assemblage. The fundamental image of the cosmos into which the matrix is set calls into question the will to system in philosophy, while the exigencies of life suggest a need to delineate affinities and inter-involvements between the elements.

As we proceed—the "we" is purely invitational—we situate the preliminary understandings and maxims somewhere between atemporal formulae and responses to immediate contingencies, folding into them a background awareness of collective issues to be addressed in the near future. The sort of assumptions and dispositions that might help us to address issues such as global warming, the failures of neoliberal economics, the excesses of the evangelical Right, the authoritarianism of the Vatican, the expansion of economic inequality, and the refusal of many constituencies to affirm the veritable minoritization of the world that today tracks the acceleration of pace and the intensification of capital. We also bear in mind how the vision of the human condition that informs this effort remains contestable. It is *how* such contests are waged that is important.

Here, then, are some assumptions, virtues, and projections in a post-Kantian mode of practical wisdom.

*The will*: The will is neither an eternal expression of suprasensible freedom, nor reducible to the determinations of efficient causality, nor the carrier of an original taint of sin. Rather, it must be decriminalized in the first instance, as part of a larger effort to overcome the culture of existential resentment that so easily grows up within it. The will is here conceived as an *emergent*, biocultural formation, which bears traces and marks of that from which it emerged but is not reducible to them. Just as life need not be devalued because it has evolved from nonlife and is now irreducible to it, and as the human species need not be devalued because it has evolved from other species, the will is not devalued because it is a partial, sensual formation installed in beings who were not predesigned to be agents of free will. As a formation it is simultaneously imperfect in shape, needed culturally, and at risk, as it were, for criminalization by those who demand universal faith in a God who bears no responsibility for the rift in being we experience.

The will, so conceived, consists of two dimensions: (a) incipient tendencies to action that well up within you as you respond to events, and (b) a limited capacity to veto or redirect some tendencies as they approach the tipping point of action. The neuroscientist Benjamin Libet, who has measured the half-second delay between the incipience of body/brain activity and its consolidation in action, suggests that the will is reducible to that nanomoment when you can veto a tendency to action under way.[19] To me, however, the will consists of both dimensions together: culturally imbued tendencies to action and a certain power to veto or redirect each as it unfolds. The will *is* two sided, as the brain is two sided, but not criminalized at its core.

Each dimension of the will is open to a degree of self-correction or modification. You adjust incipient tendencies, when reflection or the shock of new experience renders this advisable, by tactics that work upon imbued predispositions to action below the reach of the self's direct intellectual control. That is, you consciously apply tactics to yourself to help recode some of the preliminary dispositions to action below direct intellectual regulation. You also work on the capacity to exercise veto power by periodically reengaging the relation between the temporal situation you find yourself in and presumptions of practical wisdom already installed in your memory bank.

So the will is both an expression of freedom and periodically fraught with conflict without being linked to primordial guilt or lifted above organic life. It is a thing of this world in the way that thinking is, and both of its dimensions are susceptible to being worked upon tactically to some degree.

*Ethics*: In a world of becoming, in which periods of perdurance in this or that zone are occasionally punctuated by times of accelerated change, a notion of pure, universal morality, along with the idea of linear progress attached to it, requires reconfiguration. I am not talking only about whether linear progress actually occurs, but about how to proceed when a sudden, unexpected turn takes place in the trajectory you had projected forward and to which your judgments of moral principle were attuned. As an unexpected change occurs, it might be important to adjust significantly the logic of extrapolation upon which current projections of possible progress are based, or to return to work on the undergrowth of habit and disposition that has grown up like a tropical tangle in and around the principles we accept. Now ethics and politics become more intercoded and interinvolved.

Given the bumpiness of time as becoming an ethic of cultivation now assumes priority over a morality of universal law, though you may move back and forth between periods when established habits and principles suffice and those when a significant change is needed. The initial drive is to amplify, by whatever modes of cultivation available, preliminary strains of care for the earth and the vitality of life already circulating to some degree in and between most people much of the time. The idea is to fold those dispositions more actively into established patterns of desire, faith, identity, and self-interest rather than to rise to a disinterested level above the mundane worlds of desire, instrumentality, and politics.

Such dispositions have both individual and collective dimensions, as their corollaries in Kantian philosophy do too. With constituencies the quest is to fold a positive ethos into the institutions of work, investment, church, schools, consumption, corporate practice, and state policy.[20] What do you do when you encounter those without such strains? The problem here is comparable to its corollary in Kantianism, the point at which he moves either to those modes of induction or to punishment. The first move is to listen more closely than you have heretofore for aspirations and understandings by others that have escaped you and to chords of attachment you may have missed; the second is to dramatize this seed of existential attachment so that it might grow further; the third is to join forces with others to resist the most ruthless attempts to foreclose diversity or to sacrifice the future of the earth to the demands of the present.

An ethic of care is not derived; it *emerges* from a care for the world that is already there to some degree if you are lucky. And if that care is altogether lacking, logical modes of derivation are unlikely to install it. In a world of becoming it is imperative to augment preliminary strains of care with modes of sensitivity that equip you to come to terms with new twists and turns in time in creative ways.

*Periodic hesitation*: So cultivation is indispensable but not enough. In a world of becoming strategic moments (including relatively extended periods) sometimes arrive when it is pertinent to dwell in an exploratory way in the gap between a new disturbance and prior investments of habit, passion, faith, identity, progress, and political priority. In the Greek tradition those who specialized in such activities were called seers; in the religions of the Book they are often called mystics or prophets. Those who experience the world as becoming also seek to be seers periodically, in a somewhat different key. We do not listen to gods who exceed our knowledge, limited as it is. We allow multiple pressures and concerns to reverberate through us during fecund moments in the hope that a new, untimely idea, theme, or strategy will emerge for further exploration. The idea is untimely because it is not yet coordinated closely with a set of others taken for granted during that time.

How do you proceed? As an erstwhile friend has said, during a protracted present of potential metamorphosis, "it is important to ignore no signal from the emotions of whatever kind"; you also seek to absorb "the slightest instigation," as you immerse yourself in a *hypersensuous* situation in which new disturbances are absorbed experimentally and some fixed judgments begin to melt away.[21] To have previously cultivated care for a world in which such moments of accelerated change periodically arise is to prepare for such exercises in dwelling.

When things are relatively stabilized, presumptive faith in established judgment is often reasonable. Such a judgment is only presumptive, however, because stable contexts can often obscure or legitimize sources of danger and modes of suffering in need of redress. Things become most dicey, however, during periods of accelerated change in this or that zone. Now the task is to dwell with exquisite sensitivity in an emerging situation, allowing latent memories, established codes, care for being, existential worries, and emerging pressures to resonate back and forth, almost mindlessly. Out of that process a new idea, maxim, directive, or practical imperative may emerge for consideration. The next task is to subject it to experimental action to explore its consequences in the emergent context. Here judgment, creativity, and experimental action fold into each other, with each making a difference to what the other is at its best.

*Responsibility*: The task is to readjust the usual balance between attributions of responsibility to self or others for the wrongs committed and the cultivation of presumptive responsiveness to others whose ways are not readily discernible to you. The prevailing priority, the one to be adjusted, represents the primacy of the Augustinian/Kantian tradition in the Western world today. In a world of becoming it is less often the case that a simple equation can be reached between the evil that is experienced and a set of agents singularly responsible for it. On the other hand, in a world of becoming it is even more important than heretofore appreciated to cultivate presumptive responsibility—critical responsiveness—to new constituencies and demands as they surge into being. Both dimensions of responsibility are needed, but the current distribution of priority between them requires significant adjustment. This is even more true during a time when the Euro-American world—and elsewhere too—is being minoritized along more dimensions and at a faster rate than heretofore.

*Militance*: At this historical conjuncture, as it were, neither revolutionary *élan* nor liberal reform suffices. The former too often devolves into waiting for the next radical break to arrive or courts a fascist response by overreach. The latter is too inattentive to how time is often out of joint with itself; for that reason its conceptions of ethics and political action are not experimental enough.

In place of Kant's world ethical commonwealth, we pursue militant democratic assemblages that speak positively to the minoritization of the world taking place at a more rapid pace today, to the effects of systemic regional inequality, to the dangers of climate change, and to the depletion of essential resources. Anchored in no single class, faith, or generation, formation of such an assemblage involves moving back and forth between the micropolitics of media life and local involvements, the internal ventilation of faith constituencies to which we belong, confrontation of corporate leaders, investments in electoral politics, and selective participation in cross-state citizen movements. Of course there is a sharp tension between the call to dwell on occasion and the call to militance. But that tension does not take the form of a performative contradiction. Rather, oscillation between these two modalities is necessary for either to work well today. Dwelling can help to infuse militance with new ideas and strategies in a changing world.

The will as biocultural emergent, an ethic of cultivation, a world of becoming, periodic dwelling, presumptive responsiveness, untimely wagers, attachment to this world, a timely militance. (What a world!) This problematic is not a species of "relativism." Those of us who think that existential resentment is a dangerous temptation built into the human condition itself, and seek to address it in manifold ways, are hardly relativists. Those who think the current condition exacerbates that very danger are not either. It is not relativism, first, because it identifies recurrent forces to overcome in several contexts; second, because it does not automatically accept the rules and norms already operative in this or that place; third, because it solicits a broad care for the earth and the fundamental diversity of being across various traditions; and, fourth, because it often commends militant engagement with prevailing forces.

It is not "absolutism" either, since the call to dwell creatively in new situations may issue in an insight that challenges something in preexisting interpretations of God, principle, morality, causality, instrumental reason, providence, mastery, or time—and since its advocates seldom contend that they have *proven* the most basic dictates they bring into the public world. It is, rather, a set of maxims of practical wisdom, oriented to a world of becoming in which multiple force fields set on different tiers of chrono-time periodically collide or coalesce to foment a new danger, risk, or possibility. Such an assemblage of understandings, projections, and maxims tracks and displaces corollary movements in Kantian instrumental reason, practical reason, cosmopolitanism, the moral duty to assume linear progress, and a world ethical commonwealth. That is its most profound debt to Kant.

The next task is to pursue, wherever possible, relations of agonistic respect with Kantians, neo-Kantians and others who either come to acknowledge without deep resentment the relational contestability of their own philosophies and faiths *or* do a hell of a lot better job than heretofore in demonstrating their necessity. In pursuing such engagements we appeal to affinities of spiritual attachment to this world that cut across the manifold differences in doctrine. We pursue lines of spiritual connection across multiple creedal differences.

The last dimension of the maxims advanced here, then, is the element of self-reflexivity attached to them as the comparative contestability of the complex is acknowledged. We advance this perspective with a mixture of shock therapy, argument, evidence, dramatization, and modes of attachment to this world. We invite others to engage those elements. And we periodically recoil back in an invitational way upon the potential contestability of our own operational assumptions and maxims, including our image of time. That is the invitational dimension invested in such a set of maxims, a dimension already discernible in minor figures such as Jocasta, Haemon, Ismene, Eurydice, and the Messenger in the plays of Sophocles, even if absent from the major figures. Haemon, for instance, calls upon his father to relent and compromise his principles of statecraft at a critical moment in *Antigone* when time is running out. By the time Oedipus reaches Colonus, he may even present a living/dying challenge to the generalization about the major figures.[22] Does the old man finally embody something of the practical wisdom of the mature Sophocles?

# 5

## Freud's Helplessness

### Adam Phillips

"I am myself alone," Richard, Duke of Gloucester—the future Richard III—boasts toward the end of the third part of Shakespeare's *Henry VI*; and to modern ears there is the ambiguity of his being both by himself, and only himself. But he was also living in what we have come to call an enchanted world, of spirits and demons, and indeed of God. There was a limit set to how alone he could claim to be; and indeed a limit set to how much of his singularity he could claim to know. "Conscience is but a word that cowards use" (5.3.110), he suggests in a covert acknowledgment of his own cowardice, and of what we might want to think of as a disavowed part of himself (he has registered the word, but only as a word, and only as a word used by others). "I am myself alone," as a boast rather than a plaint, is a description of what Richard thinks he has been able to dispense with. It is the claim of a character one critic has called "a sardonic narcissist."[1] A narcissist is a person for whom the need for others is a conundrum. The sardonic are, in the words of the *OED*, "bitter, scornful, mocking."

Richard, we can say, implicitly but consistently ablates his own helplessness; his language about himself, in all its tortuous and subtle dissimulation, insists on his own invulnerability. Or rather, it is an attempt at invulnerability (which is always linked to sadism). So it is of interest, I think, that one of Shakespeare's three uses of the word "helpless" is in *Richard III* (the other two are in *The Comedy of Errors*); the first use of "helplessness" cited by the *OED* is 1731, and it is, I imagine, of some significance that it is a term that is first used in the eighteenth century. The word is used early in the first act by Lady Anne as she is brought the dead body of her father-in-law, King Henry VI. The man who, within the space of the play, will become Richard III, has had both her husband and his father murdered, and will eventually marry her; but at this moment in the play she is faced with the king's body, and the impending catastrophe. Looking at his wounded body, and weeping, she says: "Lo, in these windows that let forth thy life / I pour the helpless balm of my poor eyes" (1.2.13). She is helplessly crying, and her tears, the helpless balm, can't help the dead king or herself. What she can't help but express, the tears she is shedding, are no help; her helplessness is no help to her (it is, of course, a familiar topos that helplessness can't be helped, and is no help; consider Beckett's "Ill Seen, Ill Said." "She sits on, erect and rigid, in the deepening gloom. Such helplessness to move she cannot help,"[2] and the ineffectuality of tears is a continual theme). In this case her tears can neither bring him back to life nor stem the evil that is Richard.

Nothing can make her tears helpful balm; indeed helpless balm is a contradiction in terms, because if it is helpless, it is not balm at all. In this desperate scene nothing can be done; helplessness is tantamount to hopelessness. Lady Anne can refer to the king's dead body as a "holy load," but again to our modern, or rather secular, ears it could be—as such scenes more obviously are in Shakespeare's later tragedies—in miniature, a scene of what we might be tempted to call "catastrophic disillusionment," and that we have learned, from Weber, to call disenchantment—the acknowledgment that there is no (redemptive) magic.

Can we experience helplessness—can we notice that there may be such a thing as "helpless balm"—without needing to reenchant the world, that is to say, without talking of religious providentialism of one kind or another, or now of the wonders of science and technology, or indeed of art? Can we acknowledge our helplessness and do without what Leo Bersani has called "the culture of redemption"?[3] How has it come about that something so fundamental to our being as helplessness is akin for us to hopelessness? Richard, we might think, and as Freud did think, enacted a malign solution to his own helplessness. What would a benign solution be? Why, to put it as starkly as possible, does our helplessness so often tend to make us what we call bad?

In "Some Character-Types Met With in Psycho-Analytic Work" (1916), Freud uses Richard III as his example of what he calls the exceptions, those people whose neuroses, Freud writes, "were connected with some experience or suffering to which they had been subjected in their earliest childhood, one in respect of which they knew themselves to be guiltless, and which they could look upon as an unjust disadvantage. The privileges they claimed as a result of this injustice and the rebelliousness it engendered, had contributed not a little to intensify the conflicts leading to the outbreak of their neurosis."[4] The exceptions have suffered something they could do nothing about, and made a privilege of necessity: they have been inspired, so to speak, by a bad bout of helplessness.

Richard in his opening soliloquy says, as paraphrased by Freud: "Nature has done me a grievous wrong in denying me the beauty of form which wins human love. Life owes me reparation for this, and I will see that I get it. I have a right to be an exception, to disregard the scruples by which others let themselves be held back. I may do wrong myself, since wrong has been done to me."[5] The exceptions, in Freud's examples, are not the agents of their undoing; their victimhood becomes a form of entitlement. For Richard it is, in a sense, an opportunity to invent his own morality, or, at least, to exempt himself from the morality of others. "I, that am curtailed of this fair proportion," he says in the famous soliloquy, quoted by Freud,

> Cheated of feature by dissembling Nature,
> Deform'd, unfinished, sent before my time
> Into this breathing world, scarce half made up,
> And that so lamely and unfashionable,
> That dogs bark at me as I halt by them.
>
> (1.1.19–23)

Freud, as we shall see, comes back to this stark image of being sent before our time into this breathing world "scarce half made up" (Richard, Freud remarks, "is an enormous magnification of something we find in ourselves as well").[6] It is an image of Richard's original helplessness; it is a predicament inflicted. Like Lady Anne's "helpless balm" of tears, he couldn't help it happening to him, and this became, through his self-cure for it, no help for him or others. None of us choose our "feature." "We all think we have reason to reproach Nature and our destiny for congenital and infantile disadvantages," Freud writes by way of commentary on Richard's soliloquy; "we all demand reparation for early wounds to our narcissism, our self-love."[7] Freud, as we shall also see, turns this fundamental situation—what we make out of our original helplessness—into an explanation of the origins of culture; of morality, of religion, and of art. If we all have a sense of ourselves as helplessly disfigured in early childhood, we must have a picture, a sense of what it would be not to be so disfigured (the "if only" life that will inform our grievance). Something was done to us—or, as in Richard's case, it was as though something was done to us—and we were helpless. Either it couldn't be helped, or no help was available. Helpless means unprotected. It means, in this context, having an impotence foisted upon us, and organizing a life around this fact.

In this story it is clearly not part of our narcissism, part of our self-love, to be helpless in this way, or it is part of a negative narcissism, the specialness conferred by an exceptional suffering. And if it weren't for this helplessness, we would not suffer in the way we do. What is being conjured here is a counterimage of invulnerability, the opposite of help-lessness. In this equation, it is our helplessness that makes us so narcis-sistically vulnerable, that makes our self-love so precarious. The hell of the narcissist, Serge Viderman once remarked, is the tyranny of his need for the other.[8] There is helplessness, in other words, and there is the lure of self-sufficiency, of creating the illusion of being everything to oneself. And yet, of course, in a certain sense, helplessness is where we start from; or, as object-relations theorists put it, dependence is where we start from. Either way we can't get round the fact that, as Winnicott put it, all phi-losophers were once babies.

I want to consider in this paper Freud's story about helplessness with a view to making a case for it; a case for helplessness as something we shouldn't want to think of ourselves as growing out of. We can become more competent, but we shouldn't imagine that we become less helpless, the wonderful phrase "learned helplessness" reminding us, of course, that it can be learned but also unlearned.

"Moral philosophy," Charles Taylor wrote in *Sources of the Self*, "has tended to focus on what it is right to do rather than on what it is good to be, on defining the content of obligation rather than the nature of the good life ... it has no conceptual place left for the notion of the good as the object of our love or our allegiance or, as Iris Murdoch portrayed it in her work, as the privileged focus of attention or will."[9] In psychoanalysis the question, I think, has always been about what it is right to do about helplessness, rather than about helplessness as integral to the nature of the good life; or indeed as the object of our love or our allegiance. I think it should be the privileged focus of our attention, though probably not of our will. In the story I will be telling—and that Freud in some ways tells and in some ways doesn't—acknowledgment of dependence is no more of a solution to helplessness than is the injunction to pull up your socks. And this is partly because helplessness is more often than not assumed to be the problem, or what we are suffering from, rather than a pleasure, a strength, or a virtue. It is not something, as it were, that we cultivate. We do not think of development as a project in which we want to become increasingly helpless, or one in which we elaborate and sophisticate our capacity for helplessness.

One of the moral questions ushered in by psychoanalysis is this: what kind of good are the things we can't help but say, or do, or feel, or think, or desire? I want to add to this the question, what good is helplessness, a question that perhaps inevitably exercised Freud; indeed virtually everything Freud wrote was about not only what can't be helped, and what can; but also, and more interestingly, about the moral consequences of our helplessness. So what is called helplessness, and what good, if any, could it be?

In a section of Freud's early *Project for a Scientific Psychology* entitled "The Experience of Satisfaction," Freud is trying to give an account, in neuronal terms, of how the infant, the rudimentary person, manages the stimulation of appetite. It is an interesting passage not least because it shows Freud using the language of science in a way that leads him into the language of morality; in the terminology of the Project we see that once the experience of satisfaction becomes the topic, the language of neurology becomes permeable to moral preoccupations; what Freud calls "relief of tension," "discharge," turns inevitably into questions about the Good, about adequate ethical objects and "moral motives."[10] Freud is working out an account, though this is not how he would have put it, of what the philosopher Alasdair MacIntyre calls "the distinctive virtues of dependent rational animals, whose dependence, rationality and animality have to be understood in relationship to each other."[11] When Freud talks of the experience of satisfaction, he is talking in the first instance of what is conducive to survival, and in the second instance, as it were, to put it rather more ambitiously, about what might be conducive to human flourishing. Satisfaction is the word, the experience, that links what we have learned to call our desire and our obligation. And Freud, unsurprisingly, can't talk about any of this without invoking the idea of helplessness, without, indeed, making the helplessness of the human infant the heart of the matter. "Experience shows," Freud writes, keeping things as empirical as possible, that once appetite occurs,

> the first path to be followed is that leading to internal change (e.g. emotional expression, screaming, or vascular innervation). But no discharge of this kind can bring about any relief of tension, because endogenous stimuli continue to be received in spite of it. . . . Here a removal of the stimulus can only be effected by an intervention which will temporarily stop the

release of quantity in the interior of the body, and an intervention of this kind requires an alteration in the external world (e.g. the supply of nourishment or the proximity of the sexual object), and this, as a "specific action," can only be brought about in particular ways. At early stages the human organism is incapable of achieving this specific action. It is brought about by extraneous help, when the attention of an experienced person has been drawn to the child's condition by a discharge taking place along the path of internal change [e.g., by the child's screaming]. This path of discharge thus acquires an extremely important secondary function—viz., of bringing about an understanding with other people; and the original helplessness of human beings is thus the primal source of all moral motives.

When the extraneous helper has carried out the specific action in the external world on behalf of the helpless subject, the latter is in a position, by means of reflex contrivances, immediately to perform what is necessary in the interior of his body in order to remove the endogenous stimuli. This total event then constitutes an "experience of satisfaction," which has the most momentous consequences in the functional development of the individual.[12]

The picture is of the desiring "helpless subject" "filling" from within with stimuli and seeking in the first instance relief through physical expression; this reflex magic fails because the stimuli of desire keep coming. What Freud calls the release of this quantity can be effected only by an intervention by someone else, someone outside. These specific actions—put in inverted commas by Freud—can be brought about only in "particular ways," the supply of nourishment or the arrival of the sexual object depending upon the attentiveness, the quality of attention of someone else, and their generosity. The essential thing, without which the child quite literally would not survive, is provided by someone else. Desire is made viable by its recipient. It is worth noting that by equating here, in this way, the supply of nourishment with the proximity of the sexual object, Freud is making them equally urgent needs; making us wonder what happens to the sexually desiring subject, the "helpless subject," as Freud calls her, if her sexual need is not attended to, given that we know what happens to the unattended baby.

But it is, in some ways, a simple point: the helpless subject needs help. Help is not something added on afterward; it is integral. There is no position, no stage or state, before helplessness; and there is no stage before what we call help is required. But then Freud makes his remarkable statement, almost as an afterthought: "This path of discharge"—that is, the scream, the emotional expression—"thus acquires an extremely important secondary function—viz., of bringing about an understanding with other people; and the original helplessness of human beings is thus the primal source of all moral motives." The first function of the scream is an attempt at evacuation, at discharge of stimuli, which can't work; but the secondary function is as an appeal, a medium of communication or contact, between the "helpless subject" and the person looking after him. It brings about an understanding with other people presumably because it stimulates the other person to work out, to imagine what it might be that the helpless subject is in need of. It makes the other person think about what might be good for the helpless subject; and, presumably, about whether it is good for this other person to try to provide what is needed.

"The original helplessness," Freud writes, "is thus the primal source of all moral motives." Moral motives might be construed as predispositions, or reasons, or causes to pursue the good. It is a usefully ambiguous phrase in the translation; is the original helplessness of human beings the primal source of all moral motives in the infant as helpless subject, or does the infant's helplessness—and the sexual adult's helplessness—call up moral motives in the recipient? Does our original helplessness make us moral, or is morality prompted in us by the way we respond to dependent others? Freud is making a link between our original helplessness and the primal source of all moral motives; as though he is saying, without original helplessness there would be no moral motives. As though, however it works, morality is what we have invented, what has been summoned up in us, by our helplessness. Because we are originally helpless subjects—though by linking the hungry infant with the desiring sexual adult, Freud is more than intimating not merely an original helplessness, but an enduring or constitutive helplessness—we can't separate out obligation from need; we cannot help but consider what we need to do to, for and with the people upon whom we depend. To be a helpless desiring subject is to be implicated, to be enmeshed in and inextricable from, a world of moral considerations.

And among the things, of course, that this will involve us in are attempts—however forlorn or desperate or intermittent—to want to separate out desire and obligation. The way we tend to do this is to disavow helplessness. Once we keep helplessness in the picture—put it, as Freud does, in the middle of the picture—we can't dissociate appetite and morality. You can say, as Richard does, "I am myself alone" as a boast rather than as a statement of despair only if you have found a way of creating the illusion that you are not, in any way, what Freud calls "a helpless subject," a subject who needs help, who is, indeed, only a subject because he has been helped. It is original helplessness that leads us to, that makes necessary, the idea of the good.

But of course Freud doesn't say the original helplessness of human beings is thus the primal source of all good moral motives; he just says morality is bound up with helplessness. It is worth recalling Charles Taylor's comment, quoted earlier, that, "moral philosophy has tended to focus on what it is right to do rather than on what it is good to be, on defining the content of obligation rather than the nature of the good life." Morality, we might say now, is what we have tended to do with original helplessness. Morality is what we have made out of it. It is, at least in Freud's view—but these are my words, not his—our self-cure, for better and for worse, for the fact that we are helpless subjects. And the problem, we might say, is that we have tended to be—or have been tempted or inclined to be, for reasons that we will discuss—more like Richard; and by that I mean we could begin to see that much of our discontent with morality, much of our sense, when it exists, that morality is alien rather than integral, a foreign body foisted on us to deprive us of our real satisfaction, comes from the ways in which we can use morality to deny, abolish, refuse, disparage, trivialize, and punish our original helplessness. Or to put it another way, or to put it the other way round, any morality that does not affirm, desire, and value helplessness is merely punitive. Any morality that is not on the side of helplessness—that can't bear to see its pleasures and its strengths—is going to feel estranging. So we need to consider what might have to happen to original helplessness that might make it a vice rather than a virtue, a persecution rather than a boon. What would make us so averse to what is so original about us?

And, at least in the passage from the Project, Freud has a fairly obvious answer to this question: helplessness becomes persecution, is made into a problem, by being insufficiently responded to. If the hungry infant's needs are not at least recognized, not always actually met, if the sexually desiring adult finds no object attentive to his desire, helplessness becomes intolerable; something has to be done with it (it might be turned into omnipotence, say, or, better, scornful, mocking behavior, which may be the same thing). But this is the reassuring, commonsensical account, one that has been taken up in various versions of object-relations theory. It is not our nature that is the problem, but it is our parenting that can make it so. You will notice in this account that helplessness is the precondition for being helped; as an experience in itself—as in the developmental theories of psychoanalysis—it is not a good one; it is what it leads to that is of paramount importance—that is, the possibility of being understood and the generation, or evocation, of all moral motives. Helplessness is the precondition for human bonds, for exchange; you have to be a helpless subject in order to be helped, in order to be understood, in order to become a moral creature. Helplessness can make us good. And so, by the same token, if you can't experience helplessness, you are precluded from these fundamental human experiences. To get back, or to be brought back, to helplessness is to be brought back to these things. So we shouldn't underestimate, from this point of view, the conscious or unconscious desire for helplessness that must exist alongside the wish to refuse it. We could indeed think of ourselves as suffering from an incapacity for, or a refusal of, states of helplessness precisely because they reconnect us with these things that helplessness makes possible. Logically, states of helplessness are to be avoided; Freud gives us a picture both of why they might be desired, and of the risks of desiring them. For Freud it is only this helpless subject that is capable of experiences of satisfaction, those experiences he describes as having "momentous consequences in the functional development of the individual."

No helplessness, no satisfaction. Helplessness, of which it is so difficult to find a picture for the adult that is not simply terrifying, is the precondition for satisfaction. If we lose, or forget, or repress, or project, or attack this original helplessness, we quite literally lose, in Freud's terms, the real possibilities for satisfaction. We become one of the exceptions, like Richard. Without helplessness there can be no possibility of satisfaction, and without the possibility of satisfaction, there can be no aliveness, no point. Helplessness, Freud is suggesting, is the most important thing about us. And yet, as he also says, helplessness is the very thing we are prone to magic away, largely through religion, but also through art and morality (and so, by implication, by theory). If we can't bear helplessness, we can't bear satisfaction; there is a plot against helplessness, which turns out to be a plot against satisfaction. Real satisfaction, Freud implies, depends upon living without illusions, without the wishful magic of religious beliefs. The experience of satisfaction literally depends upon our living in Weber's disenchanted world; a world without omnipotence in it. "The psychical origin of religious ideas," Freud writes in *The Future of an Illusion,*

> which are given out as teachings, are not precipitates of experience or end results of thinking: they are illusions, fulfilments of the oldest, strongest and most urgent wishes of mankind. The secret of their strength lies in the strength of those wishes. As we already know, the terrifying impression of helplessness in childhood aroused the need for protection—for protection

through love—which was provided by the father; and the recognition that this helplessness lasts throughout life made it necessary to cling to the existence of a father, but this time a more powerful one. Thus the benevolent rule of a divine Providence allays our fears of the dangers of life; the establishment of a moral world-order ensures the fulfilment of the demands of justice, which have so often remained unfulfilled in human civilisation; and the prolongation of earthly existence in a future life provides the local and temporal framework in which these wish-fullfilments shall take place.[13]

This is, as many people have noticed, a rather sweeping account of religion. But more than thirty years after the Project once again everything comes out of the helplessness of childhood. Though certain things have changed. First of all "the original helplessness of human beings" has become, for the older Freud, "the terrifying impression of helplessness in childhood"; and what this helplessness produces is not the experience of satisfaction, but the illusions of religion, with its all-too-plausible morality that is merely a wishful concealment of our real helplessness faced with the dangers of reality. Now helplessness prompts us only to obscure its existence; it is not the route to satisfaction but the root to a knowing self-deception. And this deception can be made complete only by Freud's wheeling on a spurious father figure. In the Project it was at least implied that the infant's helplessness would be met by the mother, and the adult's sexual desire would be met by a satisfying real object of no specified gender; here the solution to helplessness, which is in fact an evasion of it, is a faked-up father. Freud, we might infer, now believes that the majority of modern people can only turn against helplessness, satisfaction, and truth; they take refuge in religion and its morality because—unlike that "helpless subject" of the Project—they can't bear or bear with their helplessness. It makes men invent a cartoon character, a parody, a man without disabilities, a man beyond help.

## II

There is another world, but it is in this one.
—Paul Eluard

There are, let us say, two solutions that Freud proposes to our original human helplessness, a good one and a bad one. In the good one helplessness is the precondition for satisfaction, the only way to the experience of satisfaction; and by the same token the only way to morality. In the bad one, the experience of satisfaction is replaced by the experience of feeling protected. Helplessness issues in the wish to be protected from the experience of helplessness, not to feel it too acutely. Helplessness is not recognized, so to speak, as a predisposition toward sensuous satisfaction; it is as though someone has said, "I need a drink," and the other person has replied, "It's not a drink that you really need"; or as though someone has said, "I'm hungry," and the other person has replied, "No, you're terrified." So helplessness leads to satisfaction, or to self-deception; it produces understanding between people, or misunderstanding; and, in both versions, it leads to morality, the undisclosed morality of Freud's phrase in the Project, "the primal source of all moral motives," or the morality Freud clearly despises, of the religiously consoled. It is, one might say, the difference between those who can bear being children and those who can't. Believing in religion is like believing that adulthood is the solution to childhood. Religion, for Freud, is the false solution to human helplessness—the human helplessness that Freud is keen to insist lasts throughout life; it just begins in childhood—and satisfaction is the right solution, or perhaps we should say the right outcome. But either way all of what we think of as our moral problems come out of the fact that we are helpless subjects. And helplessness, or our relation to it, is something Freud thinks we need to get right; and we do the very worst things when we get it wrong; we start doing things like believing in god, or abiding by religious teachings, or adopting preposterous moralities, or believing that we are exceptional creations rather than just another species of animal. Obviously if frustration makes us aggressive, and we turn against our own satisfaction, we are unconsciously cultivating our violence by disavowing our helplessness.

Our fundamental response to our own helplessness is to create an enchanted world, a world in which we seek protection from our helplessness, but not engagement with it. "Biologically speaking," Freud writes in his paper on Leonardo da Vinci (1910), "religiousness is to be traced to the small human child's long-drawn-out helplessness and need for help; and when at a later date he perceives how truly forlorn and weak he is when confronted with the great forces of life, he feels his condition as he did in childhood, and attempts to deny his own despondency by a regressive revival of the forces which protected his infancy."[14] It is almost as though the child underestimates his helplessness; or that what adulthood brings is a sense of just how helpless we really are. It is important, I think, that Freud nowhere suggests that we grow out of our helplessness—indeed he suggests here that we grow into it—or that it is something that can be realistically overcome. For Freud religion is a poor solution to an endemic, to a biological problem; it is because we are helpless when confronted with what he calls "the great forces of life" that we engage in this imaginative activity called religious belief. In Freud's view our helplessness doesn't diminish over time, but we become progressively more disturbed by it. So terrorized are we by it that we will seek safety rather than satisfaction, magic rather than nourishment: we seem, in Freud's view, to be the animals who are tormented by our helplessness; the animals for whom it is, or it has become, the abiding preoccupation. And in explaining this, in "Inhibitions, Symptoms and Anxiety" (1926), Freud has recourse to, or rather seems to allude to Richard III's soliloquy that he quoted in his paper on "the exceptions." "The biological factor," he writes, that makes us so prone to neuroses as a species,

> is the long period of time during which the young of the human species is in a condition of helplessness and dependence. Its intra-uterine existence seems to be short in comparison with that of most animals, and it is sent into the world in a less finished state. As a result the influence of the real external world upon it is intensified and early differentiation between the ego and the id is promoted. Moreover, the dangers of the external world have a greater importance for it, so that the value of the object which alone can protect it against them and take the place of its former intra-uterine life is enormously enhanced. The biological factor, then, establishes the earliest situations of danger and creates the need to be loved which will accompany the child through the rest of its life.[15]

Richard describes himself as "Deform'd, unfinished, sent before my time / Into this breathing world"; Freud's human animal is too briefly in the womb, and "sent into the world in a less finished state." And Freud, of course, intimates that there may be a sense in which we are all deformed by our protracted helplessness and dependence. This left Richard with a grudge, and a supposed entitlement to avenge himself, at least in Freud's view. What is the grudge? what form does the revenge of the human animal take? The value of the object that can protect it is "enormously enhanced," and the need to be loved will accompany the child through the rest of his life. What sense does it make to think of ourselves as deformed by our dependence? What picture of ourselves might we have if we were not so deformed? The long period of time that the human species is "in a condition of helplessness and dependence" leads to what we might call an idealization of the protecting object (i.e., an unrealistic apprehension of it), and a cravenness, an enslavement to being loved (Richard we might think of as idealizing himself, and enslaved to being hated). This helplessness, which accompanies us through life, calls up in us the wish for a protective object that distorts perception, and a terror of loss of love. In other words at its best, under one kind of description, our helplessness involves us with others, weaves us into the human community; but at its worst, under another description, it makes us abject and infinitely exploitable creatures (we will do anything for love and protection). It is as though our original helplessness, the defining feature of ourselves throughout our lives, at worst corrupts us, and at best makes us morally pragmatic rather than morally principled. If we live in a condition of dependence and helplessness, and if what this leads us into is spurious self-deceiving religious consolation, or the secular equivalents, how could we not end up hating our (and other people's) helplessness? how could we not end up thinking the absurd paradoxical thought that the very thing that makes us what we are ruins our lives? This, at least, is one of the places that Freud leaves us. The catastrophe of being a human being is that we are irredeemably helpless creatures. And that means, helpless to do anything about our helplessness. We are essentially helpless, and that is the very thing that makes our lives impossible. It is a mark of our resistance to our helplessness that we deem the ineluctable to be catastrophic, as though anything we don't make for ourselves is bad for us.

## III

> ... that innate incompetence, which Hooker
> had called "This our imbecillite" ...
> —Geoffrey Hill, "The Eloquence of Sober Truth"

Freud is making a distinction—though over time it is a distinction that gets lost—that at first sight seems rather crude; our original helplessness either leads, optimally as it were, to the experience of satisfaction or to the experience of religious belief, to essential sensual satisfactions of appetite, or to magic. But in making this distinction, by setting up these "contraries" (to use Blake's term), Freud is opposing protection to satisfaction. Our helplessness—that is, the helplessness of appetite—can lead us in the direction of the experience of satisfaction provided by attentive others; or our helplessness—our helplessness in the face of the dangers of the external world and the great forces of life, among which are appetite—can lead us in the direction of protection from unreal figures and magical beliefs. But this is not simply a version of our need for safety contending with our need for excitement. Freud is saying that our helplessness can force us into sacrificing, sacrificing by displacing, the experience of satisfaction; which for Freud, in his early formulation in the Project, is itself a fundamental precondition for psychic and physical survival. The experience of helplessness, in other words, can make us sacrifice our lives, can lure us into a nihilistic pact; if you give up on the experience of satisfaction, you can be protected. Of course, we don't want it to be as stark as this; we can, as we say, have both; and yet Freud, I think, is showing us something harsher: that there is something about our experience of helplessness—that there is, perhaps, something about the way helplessness is described, is presented in our culture, at this time—that makes us want to give up on our desire, and for something that can supposedly reassure us, but not fundamentally satisfy us. As though the quest for the experience of satisfaction leaves us unprotected; and we will give up on satisfaction in order not to face this.

Something about our helplessness, or its descriptions, precipitates us into a delirium of compulsive protection seeking, as though we believe we can be protected from the great forces of life. In Freud's view we are the animals who are uniquely vulnerable, oversensitive one might say, to what living entails; we are sent into the world in a "less finished state" than other animals, the dangers of the external world have a greater importance for us, and we require barriers between ourselves and our desire. As Freud puts it, "early differentiation between the ego and the id is promoted." Is this special pleading? In this account are we, rather like Richard, the exceptions, but in this case the exceptions in the whole of nature? It is not, of course, difficult to see why Freud, at the time and place of writing, should feel his own and other people's unsuitedness, ill-fittingness to the world they found themselves in. Protection would be privileged over satisfaction, for Jews and many others, in Europe in the twenties and thirties; fundamental human satisfactions would feel increasingly remote. There would be what we might think of, in short, as new forms of helplessness, new pictures and experiences of helplessness—political, economic, and, in Freud's language, psychic. And these very experiences could lead to radical redescriptions of the helplessness of infancy—of what the consequences of helplessness can be. Helplessness could only become horrifying; it might be increasingly difficult to find—and perhaps it has always been difficult to find—inspiring pictures of helplessness, or accounts of helplessness that would make it sound, in Charles Taylor's words, like something "good to be," an object of our "love" or our "allegiance." No one says things now like "he's a wonderfully helpless man," or "there was something impressively helpless about the way she dealt with that."

And yet Freud, I think, was beginning, in the Project, to make the case, or a case, for helplessness; for the fact that without the experience of helplessness there could be no possibility of the experience of satisfaction. Without the possibility of the experience of satisfaction life was futile. Indeed one way of seeing what Freud thinks of as the magic of religious belief is that it contains a long-term wish within it: that one day I will feel sufficiently protected so I will be able to get back to, to start up again, the quest for the experience of satisfaction. I will be able to sort out my helplessness—secure myself from it—so the life of appetite can begin again. But in the meantime I am going to have to believe in God. The fantasy of ultimate redemption, at least in psychoanalytic terms, might mean a disguised picture of when I can allow myself to start desiring again, in the belief my desire will be met. In the belief that there will be nothing self-destructive, nothing self-endangering about my desire (the self-endangering experience that we should remember, can be self-destructive only in retrospect). In Freud's picture of appetite, both sexual and other, in the Project, to desire, to be hungry or sexually alive, is to risk one's life. An education in helplessness, if such a thing were possible—you might say we are always already overeducated in helplessness—would be an education in mortal risk. Clearly a lot depends on what we feel about helplessness, on what, to put it as confoundingly as possible, can be done about it, or with it. Could we, for example, in the pragmatist way—which is not often the psychoanalytic way—use it to get from A to B, without the very idea of using it becoming a denial of its nature? What—to use William James's language—might helplessness be good for? In the Project, after all, Freud was telling us that helplessness was good for the experience of satisfaction. That that is what it was good for, and that that was what it was originally meant for.

Of course it seems odd to instrumentalize helplessness; but perhaps this is something we can't help but do, and this, in itself, may be instructive. Freud says, in the instances I have cited, that helplessness is good for the experience of satisfaction, and good for making us magical (i.e., religious); it makes us enchant the world with supernatural agents and forces. But Freud is also saying that helplessness never stops; that in fact we get progressively more helpless. So we might say helplessness is only temporarily transformed by satisfaction, but that religious belief creates the illusion that it is permanently transformed (we are helpless, but our helplessness is grounded by help, the infinite help of God). Developmental theory fobs us off with the sense that helplessness can be progressively transformed. What Freud is urging us to imagine is what our lives would be like if we lived as if—at some level, in some way—we were all helpless all the time; and that this helplessness was the very thing we sought to conceal, or obscure, or deny; that our helplessness has no solution, and there is no holding it at bay or growing out of it; that prior to sexuality, or aggression, or dependence there is helplessness as a way of life, a way of being; that we are helplessly desiring creatures. That our original (in both senses) and fundamental helplessness could be recognized, acknowledged, and even to some extent met—indeed this is where object-relations theory, attachment, relational psychoanalysis, and other stories about dependence come in—but in all these pictures our helplessness is not temporarily relieved, or assuaged, or removed; it is just that so-called relationship is one of the things we do with, or about, our abiding and presiding helplessness. And if helpless is simply what we are—there is no other way of being, no way of being anything else—how and why have we made it such a problem? Or to put it the other way round, why have we not seen it, in Taylor's words, as something good to be, not just bad to be?

Indeed, we have, in a sense, defined the Good as that which relieves us of our helplessness, or makes it bearable. So when Charles Taylor asks, in *Sources of the Self*, "What is it about the [human] subject that makes him recognise and love the good?"[16] the answer could be, the subject's original helplessness; our helplessness makes the good, and the idea of the good, integral, essential to our survival and well-being, whether the good is God, or the mother, or ideological commitments, or moral convictions, or, as we say, the meeting of needs. There is an obvious religious point here, which has broader implications. "Fallen man was utterly helpless," Taylor writes, "and could do nothing by himself. The point of harping on the helplessness and depravity of mankind was to throw into the starkest relief the power and mercy of God, who could bring about a salvation which was utterly beyond human power and, what is more, still wanted to rescue his unworthy creature beyond all considerations of justice."[17] Our helplessness, then, in this usefully generalizable description—for God here we could put the mother, the parental couple, the saving ideology, and so on—is the precondition for our idea of the good; one could almost say it makes the idea of the good possible, even plausible. Without our original helplessness such a thing would never have occurred to us.

But there is a familiar moral move here that is worth noting: helplessness has to be characterized as "depravity" (to use Taylor's term), or lack, or weakness, to make the good look good. It is reminiscent of Nietzsche's striking point in *On the Genealogy of Morals*, when he writes of our "will to establish an ideal—that of the 'holy God'—and to feel the palpable certainty of [our] unworthiness with respect to that ideal."[18] What if our casting of helplessness as weakness, as a form of depravity, has produced spurious forms of strength of the good as that which relieves us of, or even conceals, our helplessness? When Freud implies in the Project that we can satisfy each other but not save each other, we might take him to be saying: satisfaction, cumulatively experienced—or even satisfaction risked—can make helplessness a strength. Indeed if, as Freud does, you make the experience of satisfaction, and the frustration that leads to it, into the constitutive human experiences, then helplessness—that is, the helplessness of appetite, of desiring—is our fundamental strength. Without it there is no frustration and no possibility of the experience of satisfaction. Is helplessness only a good if and when we are helped?

So if the Good were to be formulated not as an overcoming of helplessness, or a compensation for it, or a denial of it; if the good, in other words were not itself believed to be without helplessness (as God is), what would it be like? And my invoking of Nietzsche's *Genealogy of Morals* is, of course, a warning in the background because the risk is, in promoting helplessness as a virtue, as an original strength, that we might simply be elaborating what Nietzsche calls, in his distasteful and exhilarating way, the "slave revolt in morals," the tyranny of the weak over the strong, what he thinks of as the corrupt and corrupting "intelligence introduced by the powerless." Can we, in his terms, which I think are useful here, "approach the problems of morality in high spirits," by making the case for helplessness, or does considering the sheer scale of our helplessness make us feel weaker; as though even talking about helplessness could drive us to despair, or to sentimentality, or to object-relations, or to drink?[19] Why, in short, does helplessness make us think of consolation rather than inspiration? Why is it associated in our minds with being tortured rather than being high spirited, with being desperate rather than being available, with S & M rather than abandon?

"You can either resent the way life is ordained, or be intrigued by it," the critic Denis Donoghue wrote.[20] I think we are inclined—and perhaps even encouraged, even educated—to resent our helplessness, to be frightened of it rather than intrigued by it. There is, of course, one sense in which our helplessness is not ordained: we are not born helpless; we become helpless. For the first years of our life it never occurs to us, so to speak, that helpless is what we are. Helplessness is something that, over time, we learn about ourselves; it becomes the word we might use for certain experiences. And then, it seems, once we have got the idea, helplessness becomes the thing, the condition, that we are always trying to do something about. Because we seek to relieve our helplessness, we think of it as something that we suffer, or suffer from. But what if we thought of ourselves as getting progressively more helpless as we got older, and helplessness as something we grow into, partly by becoming aware of it? And in order to do this we might have to broaden the analogy-horizon. After all, we don't think of ourselves as relieving our need for oxygen by breathing, whereas we do think of ourselves as relieving our hunger with food. We can't satisfy our need not to be helpless.

So by way of conclusion I have two suggestions, both, I hope, in the spirit of Freud's "The Creative Writer and Daydreaming": "The true ars poetica," Freud writes in this paper, "lies in the technique by which [the artist] overcomes our repulsion, which certainly has to do with the barriers that arise between each single ego and the others."[21] I take it that there is something about our helplessness—the pictures we have of ourselves as helpless—that we find repulsive; and that the barriers that arise between each single ego and the others is, in part, at least, a consequence of our disavowal of our original helplessness that is the thing we have most originally in common with each other, such that the acknowledgment of this helplessness in common makes all such barriers between us seem wildly unrealistic. My two suggestions are, first, that any psychoanalysis that privileges knowing over being, insight over experience, narrative over incoherence diminishes if not actually forecloses our real acknowledgment of helplessness. If we thought of our helplessness as like a figure inside us—as, to use Hilary Putnam's phrase from another context, "a being who breaks my categories"[22] in a way we were willing to risk—not only would we be trying to overcome it, but, as Freud intimates in the Project, we would be trying to sustain it. We would think of our helplessness as sustaining us, there being nothing else, ultimately, that could do so. It would not simply be one of the best things about us, but it would be the thing, the condition, without which we could not be who we are. In this picture, there is no version of ourselves that is not helpless, even if, in different areas of our lives, there are different forms of helplessness; we are helpless in different ways.

And my second suggestion, following on from this, is that helplessness leads us too automatically into talking about dependence (and now, of course, attachment). When I was working on a ward for dying children, a mother once said to me, pointing to her daughter, "We are relying on her now." After all the doctors, after all the help that is available, we are ultimately helplessly dependent on our own bodies, on their sustaining vitality or lack of it. No other body is available but the body we happen to be. There can be virtues in necessity. And in the fact that there are necessities.

Helplessness may be another word for disenchantment. Or perhaps it is our helplessness that we are left with after the disenchantment of the world. Freud, of course, was not saying, was never saying, as some of his crudest commentators suggested, that we must choose between religion and sex; or between religious belief and sexual satisfaction (Freud was not interested in sexual satisfaction; he was interested only in its possibility, in the sustaining of desire). He was saying, among many other things, that religion can be used as a refuge from sex, something that many people had said different ways before him. But more interestingly he was saying that what Weber called the disenchantment of the world—a wholehearted secularization, a world without omnipotence, omniscience, or invulnerability—was the precondition for the possibility of instinctual satisfaction, and the moralities that come with it. As if the enchantment of the world begins to look like an unwillingness to even consider such things; that morality and the possibilities of satisfaction can be spoken about only in a language without God (or gods) in it. As though only recently have we begun to be able to think at all about sex and morality. Our moral lives, which are our erotic lives in Freud's view—sociability as appetite, not Providential design—start with and from our helplessness. Anything that serves to distract us from our original helplessness—magic of all kinds—distracts us from our very being. It stops us getting started.

# 6

❦

# A Secular Wonder

*Paolo Costa*

## At Home in the World?

What is it like to be an earthly, worldly, "secular" creature? Is it a blessing or is it a curse? The debate, as Hume noted long ago in his essay "Of the Dignity or Meanness of Human Nature," is perennial.

In a renowned passage in *The Birth of Tragedy*, Nietzsche evoked the pessimism of Greek popular wisdom, quoting the answer that, according to a legend (told by, among others, Aristotle in a fragment of his lost dialogue *Eudemos*), the "wise" Silenus reluctantly gave to King Midas's insolent curiosity and pressing desire to know "what is the best and most excellent thing for human beings":

> Wretched, ephemeral race, children of chance and tribulation, why do you force me to tell you the very thing which it would be most profitable for you *not* to hear? The very best thing is utterly beyond your reach not to have been born, not to *be*, to be *nothing*. However, the second best thing for you is to die soon.[1]

"Better not to have been born"—a resolute judgment on the misery of being alive that recurs often in ancient literature, from Homer (*Iliad*, bk. 17, lines 445–57) to Sophocles (*Oedipus at Colonus*, lines 1224–27): human life is a camouflaged capital sentence, a nasty trick played on human beings by the gods.

The same kind of glum wisdom resurfaced many centuries later in the ruminations of an Italian poet greatly admired by Nietzsche—Giacomo Leopardi—who mischievously delighted in turning upside down Pope's idea that "whatever is, is RIGHT." Leopardi's tirade is immortally sculpted in a celebrated chain of ruthless sentences:

> All is evil. That is, everything that is is evil; that each thing should exist is evil; each thing exists for the purpose of evil; existence is an evil and is

ordered for evil, the purpose of the universe is evil; order and the state, the laws, the natural progress of the universe are nothing but evil, nor are they directed towards anything but evil.[2]

But how can one get along with such a gloomy view of existence? How can one respond decently, as a moral being, to this picture of the human condition, which reminds one of the Gnostic rejection of the world?

A possible answer could be: resistance, effort, control. It is indeed noteworthy that the early twentieth century—especially in the German context—was the cradle of a modern variant of Stoicism that reacted to the dismay engendered by the new dispiriting vision of reality, apparently fostered by modern science, by rejecting any kind of comfortable consolation in the name of the primacy of reason and of a desolate lucidity. Exemplary spokesmen of this position were Sigmund Freud and Max Weber, who acted as mentors of a whole generation of austere German intellectuals. This attitude is captured perfectly in a passage in Weber's biography, written by his widow, Marianne:

> One day, when Weber was asked what his scholarship meant to him, he replied: "I want to see how much I can stand." What did he mean by that? Perhaps that he regarded it as his task to endure the *antinomies* of existence and, further, to exert to the utmost his freedom from illusions and yet to keep his ideals inviolate and preserve his ability to devote himself to them.[3]

But is there really no alternative to this well-known constellation of ideas, which combines the early modern passion for "freedom from illusions" with a stiffness toward reality that borders on disembodied indifference or insensitiveness? To what extent can the modern "buffered" self—to use Charles Taylor's well-chosen phrase[4]—only *stand up* against an objectified world (the *Gegen-stand*) and never be *in* or *open* to it? Is it true that the only alternative to an unending oscillation between dejection and endurance is a form of self-deceit, that is, surrender to intellectual and emotional maturity so as to fall back into a world of fables and lies?

Periodically, within the history of Western culture, the secular, profane dimension of existence has been portrayed, though by a minority, not as a wasteland, but as an empowering "space of presence and appearance" (in Arendt's phenomenological jargon).[5] In the classical world, this affirmative stance was mythically expressed by Odysseus's refusal to marry Calypso, the goddess who offers him immortality in return for his love.

According to Martha Nussbaum, Odysseus rejects her proposal because of the "shapelessness of the life Calypso offers." In the end,

> he is choosing the whole human package: mortal life, dangerous voyage, imperfect mortal aging woman. He is choosing, quite simply, what is his: his own history, the form of a human life and the possibilities of excellence, love, and achievement that inhabit that form.[6]

Yet, as a matter of fact, when we struggle to assess the value of a purely "secular," immanent view of existence, we do not restrict ourselves to answering affirmatively Taylor's amusingly dubbed Peggy Lee Question ("Is that all there is?").[7] Rather we articulate or opt for a specific picture of our relation with the world or with the totality of being. Here the repertoire of images is large, but not unlimited. Thus we have to clarify what such a bond looks like: whether we are "balanced" with the world around or not; "attuned" to it or not; "rooted" in it; "reconciled" with it; or, on the contrary, "weighed down" or "overburdened" by the world; "sick" of or "nauseated" by it; "indifferent" to, "terrified" by, "thrown" into, or "imprisoned" in it. Perhaps, however, the most pregnant image is the one that wavers between the two poles of familiarity and homelessness: do we feel "at home" in the world, or not?

During the last two centuries, this question has generally been answered with an appeal to disengagedness and objectivity, that is, to an indisputable "scientific" representation of the whole of "reality." The world then seems to serve as a justification, through the "facts of the matter," of the superiority of optimism or pessimism. On the contrary, however, I want in this essay to show that, in order to argue convincingly about our basic attitudes toward the world and the reasons pro or con for them (if we have any reasons at all), we do not have to insist on disinterest; we need, rather, to move toward primary and apparently objectless emotions, of the kind that are called *Stimmungen* in German. In other words, we need an epistemology of basic existential moods.[8]

In particular, I shall examine, out of many possible, one basic attitude toward the world—the readiness to be surprised or amazed by things or events—on the hypothesis that one of the preconditions for feeling at home in the world has to be sought exactly here.

But upon what does this disposition to be amazed depend? Is it somehow warranted or is it only the side effect of an irrational vitality—a sort of by-product of a Nietzschean/Zarathustrian "Ja sagen"—if not a plain self-deceit? In other words: can one be "enchanted" for the right reasons? Has the proclivity or willingness to be surprised, to expose oneself trustingly to the expressive power of the world, to its superabundance of meaning, a special connection with a confident attitude toward life? And does the ultimate meaning of the idea of "secular enchantment" lie just here?

## Feeling the Whole

"Feeling (or sensing) the world," "feeling the whole," "All-feeling": it is unclear whether such expressions mean anything at all or are, rather, merely redundant synonyms for "feeling/sensing" that, in the end, complicate more than simplify things.

In general, the idea of a sentimental link with the totality of things tends to crop up in two different, but not unrelated fields of experience: religion and art. More specifically, talk of such limit-experiences is common in Romantic aesthetics and in mystical literature. To see the point, one has only to think of Freud's treatment of "oceanic feelings" in the first chapter of *Civilization and Its Discontents*, where he begins his analysis by summing up some thoughts outlined by Romain Rolland in a private letter a couple of years before the publication of the book:

> The true source of religious sentiments [according to Rolland] consists in a peculiar feeling, which he himself is never without, which he finds confirmed by many others, and which he may suppose is present in millions of people. It is a feeling which he would like to call a sensation of "eternity," a feeling as of something limitless, unbounded—as it were, "oceanic." This feeling . . . is a purely subjective fact, not an article of faith; it brings with it no assurance of personal immortality, but it is the source of the religious energy which is seized upon by the various Churches and religious systems, directed by them into particular channels, and doubtless also exhausted by them. One may, he thinks, rightly call oneself religious on the ground of this oceanic feeling alone, even if one rejects every belief and every illusion.[9]

Needless to say, Freud was deeply skeptical about the real nature of this feeling of "an indissoluble bond, of being one with the external world as a whole,"[10] and above all about its actual "grip" on the external world. Rolland, too, for his part, was more than willing to acknowledge its "subjective" character, around which, however, Freud was keen to superadd an epistemological framework strongly marked by the "inner-outer distinction."[11] According to this view, "normally, there is nothing of which we are more certain than the feeling of our own self, of our ego," while "towards outside . . . the ego seems to maintain clear and sharp lines of demarcation."[12] Vice versa, in an "oceanic" or "all" feeling, everything is, as it were, in the foreground, and nothing in the background; all boundaries melt into the air, including the immaterial frontier that separates the ego from the external world, and one is exposed to the sheer infinity and excessiveness of experience. (There is a clear affinity here with the aesthetic experience of the sublime, where "the mind is so entirely filled with its object, that it cannot entertain any other.")[13] That is why, according to Freud, such "strange" feelings demand a "psycho-analytic—that is, a genetic—explanation."[14] But do they?

Philosophically speaking, what is striking about these kinds of feelings is that they seem to possess a sort of intentionality whose content, however, is an indeterminate one: that is, the totality of things. Hence, on the one hand, they draw attention to the very fact of "feeling" as such and, on the other hand, awaken us to the fundamental meaning of our existence, as when we ask "grammatically" complex philosophical questions such as the ones about the worth or worthlessness of nature or life itself. These are not meaningless questions—after all, we can easily understand them—but the answers to them cannot be a disembodied "argument"; they consist, rather, of a more or less vague emotion.

There seems to be something incongruous in this way of putting the matter. At first sight, the oceanic feeling looks like a limit-experience, an extraordinary event. But, at a deeper philosophical level, such occurrences, even though they push us, as it were, to the boundaries of our lived experience, do not seem exceptional at all: they are, on the contrary, minimal rather than ordinary. In fact, we have a word to indicate the less conspicuous embodiments of these "global" or intentionally indefinite feelings: moods. By "mood" I mean a fundamental emotional attunement, a *Stimmung*, a basic disposition to be affected in this or that way by the world around us. I am talking here, of course, of the receptive facet of humans' way of being, the very ordinary fact that things always "matter" in some way or other to us, and that we cannot help but be affected by things as if we were immersed in a sort of bubble of meaningfulness or, better, in an atmosphere of significance and import that we do not create from scratch but are absorbed by. The metaphor of the atmosphere should suggest not only the image of a global container, but also that of a rhythm of breathing and of a light refraction to which a living being must attune or adjust herself.

"Receptivity," however, does not mean mere passivity. With a bit of irony, sensibility or, better, sensitivity, it could be described as a sort of "borderline" phenomenon that cuts across the dividing line between activity and passivity. For it is hardly deniable that sensations, in all their motley appearances, are a form of exploration, a way of orienting oneself, of finding one's way around in the inner and outer environment by picking up more or less useful, but in any case significant, pieces of information ("affordances," to use J. J. Gibson's term of art) in view of possible actions. On the other hand, it is equally true that sensory activities are a form of affectivity, insofar as a distinctive aspect of many feelings is that they come about through receptors that, as it were, open up to the world and its fluxes by exposing themselves to their impingement or, to employ another trite metaphor, their onslaught. But the simple fact of possessing some "irritable receptors" does not make us into inert repositories of the "irritating" force of matter. After all, it is no accident that sentient beings are also motile organisms and, generally, agents. With a predictable image, the twofold bent of sense-experience—expansive and retentive—might be described as a beam of light on reality, made possible by a skillful play of refractions. Therefore, in the endeavor to understand what is going on in a sense-experience, it is philosophically unnecessary to assume the existence of an inner and independent source of luminosity (the absolute "spontaneity"), or, vice versa, to ascribe the whole dynamism to the causal powers of external reality.[15]

In this light, to sense or to feel amounts essentially to disclosing, while sensitivity appears as a necessary condition for having a world (*Welt*) or, at least, a relatively stable life-world or environment (*Umwelt* or *Lebenswelt*) that, more than being in front of us, surrounds us with its diffuse significance or import for our lives.

Given these premises, it can be plausibly argued that the experience of having-a-world has its roots not in a head-on and focused relationship with a clear-cut object, but in the emergence of a bubble of significance that, for a sentient being, plays the same role that is played by the atmosphere with regard to the earth: it creates, that is, special conditions of life where existentially crucial distinctions between inside and outside are drawn. Along with this world-disclosure comes also that middle ground—the lived body—which makes possible the experience of "feeling" (well, bad, angry, serene, excited, etc.). We are well and we feel well; we are serene and we feel serene; we are excited and we feel excited. All these familiar expressions could also be turned into more abstruse formulae such as: "I have (or find myself in) a benign world"; "I have (or find myself in) an exciting world."[16] Affectedness, situatedness, or *Befindlichkeit* (to use the original Heideggerian term), not only "colors" the world in which we live, but *is*, in a strong sense, the world in which we live. It is not in front of us as something we can manipulate or dispose of, but around us: we are immersed in moods as we are exposed and "attuned" to the weather.[17]

So far, I have not done much more than paraphrase Heidegger's famous treatment of the subject in section 29 of *Being and Time*.[18] Yet even such a cursory and sweeping overview of the matter can be of help in outlining the issue at stake, which could be summed up along these lines: our ordinary commerce with the world is underpinned by some elusive emotional states whose nature is, at a first glance, rather enigmatic. Not being correlated to a specific object or intentional reference, moods may be seen as a radically subjective phenomenon and, therefore, in ultimate analysis, impervious to reasons, if by "reasons" we mean normative nexuses that are liable to be right or wrong (in some cases, even good or bad). But if this were the case, nothing of worth could be said about the world relevance of moods. An existential affective attunement would be nothing but the consciousness side of a chemical-physical state of the organism, produced by internal factors; it would be not a way of feeling the world, but just an empty self-feeling, a mysterious epiphenomenon without any epistemological relevance: a contingent side effect of the way we are or evolution has made us.

But if we want to remain faithful to Heidegger's insight, we must back away from this style of reasoning that concocts an ontological landscape where everything seems to turn on crossing boundaries or thresholds, moving from inner to outer worlds (and vice versa), devising solutions to (mostly imaginary) dualisms, and the like. On the contrary, we must recognize that, for a living being, moods (i.e., primary affectedness) are the precondition for having a world. They are, as it were, the basic level of reality (the phenomenological analogon of the empiricist "brute fact"). Otherwise put, a purely objective, totally anesthetized reality is only another avatar of the modern epistemological Myth of the Given and a fully de-realized, evanescent, weightless one. For reality's gravity—its "resistance"—discloses itself primarily in the arousal of bodily alertness and in the simultaneous emergence of an oriented field of significance, rather than through a disengaged view of the world.

Going back to the start, one might argue that, in a sense, moods, too, are "*oceanic* feelings," to the extent that they are effusive, and their "grip" on reality is always inchoate, vague, and global. (We could also class them as *ordinary* oceanic feelings.) In this sense, they are also "ecstatic" feelings, as they go beyond (or behind) the distinction (and confrontation) between subject and object. Hence they are everything but self-enclosed mental states and work, rather, as affective starting points, sources of disclosure, insofar as they are a primary way of putting things in perspective, of experiencing them as bearers of significance and relevance for one's own life. But of an enigmatic relevance. In fact, moods operate as generic "signals" or "guides,"[19] because they point "in an ostensive mode" without fixing the frame or the end point of the interrogation,[20] and so they awake and draw our attention, enabling us to focus it, orienting the senses, and sometimes energizing them (and sometimes de-energizing them). That is why moods can also be depicted as objectless emotions, because they point in an indeterminate direction and operate for the respondent as open-ended signals. As a consequence, they help to fuel a process of reflective and "affective escalation," without which no explorative attitude toward the world would be possible. Indeed, they must be understood as conditions of articulation, insofar as they are the framework of all frameworks: the global frame.

In particular, they are the background against which the particular emotions are the foreground. Or, to use another image, moods are like the halo of emotions, which explains how, for example, one can be confronted with the same affective phenomenon under two different guises: as a mood or as an emotion. This regularly happens with fear, joy, surprise, sadness, boredom. One can be fearful about something or merely feeling fearful, and the difference, here, is the same as that which obtains between thinking hard and letting your mind wander or going somewhere and strolling around.

Understanding this simple fact should foster a (taxonomically) more relaxed approach toward the universe of emotions, passions, sensations, feelings. There is actually a continuum among the different affective phenomena, ranging from the punctual, reflexlike reactions, to the enframing power of moods. And it is the existence of this continuum that makes the affective escalations such a common feature of our ordinary experience. We can begin by being startled by thunder and end up in a general mood of astonishment. Or we can start off with a fearful mood and end by being startled by any sound in the surrounding ambience. Of course, there is no mystery here: it is only the nature of our life-world that makes a lasting dialectic between object and world not just possible but common. As Stanley Cavell once put it, "Sense-experience is to objects what moods are to the world."[21]

In conclusion, although moods, in all their variety, cannot be interpreted as competing judgments about the world in its entirety (understood as a clear-cut object), each of them helps to disclose one of the many faces of the world by attuning us to it and making it resonate with our receptivity. Their changing nature is a crucial source of basic knowledge about the world, especially about its "staggering diversity."[22] We learn a lot about the world kaleidoscope just by taking moods seriously and, with their aid, peeking into ordinary life's plasticity and *Doppelbödigkeit*.[23] Indeed, the dynamism of moods provides an excellent basis for overcoming the veil of any flattening gaze and building up a multidimensional picture of the world, mindful of all its emotional upheavals. This is a landscape worthy of study and exploration.

## A Phenomenological Sketch of Wonder

In the world's intricacy and density to which alternation of moods accustoms us, a space is naturally opened for wonder, that is, for a stance toward things that is intentional without being appropriative, the sort of attitude that Francis Bacon defined, with a marvelous linguistic intuition, as "broken knowledge,"[24] a suspended gesture that makes room for an experience of sheer presence.

Although wonder—in virtue of its world-openness and world-relatedness—can be easily identified as a fundamental mood, it is not as such a self-evident experience. So how are we to understand it? All the various definitions of "wonder" converge around a constellation of ideas. According to John Onians, for example, "to feel 'wonder' is to acknowledge the impact of an extraordinary sensory experience."[25] Similarly, Kelly Bulkeley speaks of wonder as "the feeling excited by an encounter with something novel and unexpected, something that strikes a person as intensely real, true, and/or beautiful."[26] Both these formulations are reminiscent of Descartes's influential definition: "wonder is a sudden surprise of the soul which makes it tend to consider attentively those objects which seem to it rare and extraordinary."[27]

To wonder, then, is to feel a sense of bodily and mental thrill as a result of a sensory (or imaginative or mnemonic) encounter (or failed, but awaited encounter) with something in our world that appears new and is unpredicted. The surprise helps to draw and focus the subject's attention by triggering a life-enhancing physical and intellectual arousal. In the feeling of wonder, thus, one undergoes a chain of interrelated events: (a) a striking experience; (b) a bodily and mental response; (c) a form of empowerment that can lead to a new action or a novel perspective on things. A crucial feature of the wondering attitude is, then, the "quality of attention."[28] In wondering, the world, as it were, manages to take up, to absorb, the individual's attention so that she is, in a way, sucked and "embedded" into it. And all of this takes place with a sigh of awe or dread, depending on whether the source of wonder appears as life empowering or life menacing. In this sense, wonder is always characterized by a "suspension" or "systole of the heart" (*systole cordis*), to adopt Albertus Magnus's poetic image.[29] And it is an open, outward-directed, disclosive, and, in Ronald Hepburn's words, "other-acknowledging" feeling.[30] Furthermore, the wonder experience seems to have a privileged link with the new (the human capacity to *begin*) and the uncertainty that comes with that—a fact that drove Descartes to speak of wonder (*l'admiration*) as "la première de toutes les passions":

> When the first encounter with some object surprises us, and we judge it to be new, or very different from what we knew in the past or what we supposed it was going to be, this makes us wonder and be astonished at it. And since this can happen before we know in the least whether this object is suitable to us or not, it seems to me that Wonder is the first of all passions. It has no opposite, because if the object presented has nothing in it that surprises us, we are not in the least moved by it and regard it without passion.[31]

Of course, we are speaking of wonder, but we could have easily employed another term: astonishment, amazement, surprise, awe, stupor, fascination, curiosity, or perhaps even "startlement" (I am looking, without success, for an English word that communicates the idea of the "condition of being startled," because the word "startle" conveys so intensely the almost physical shock of the experience). And, truly, any attempt to rivet unequivocally the historically shifting semantic field that encompasses such bordering notions is bound to fall short. The truth is that they all belong to a dynamic conceptual constellation where they periodically exchange their places and, along with that, alter their semantic halo. So, depending on the predominant semantic force field, curiosity may appear as a sober passion and wonder as an ephemeral one, or vice versa; awe may come first, and curiosity last, and vice versa; astonishment may look vulgar, and wonder respectable, and vice versa, and so forth.[32] For the problem is not wonder (or curiosity or awe) as such, but the nature and significance of our stance toward the world when we eagerly respond to its pull and give ourselves over to it.

Phenomenologically, the experience of wonder may be subdivided into a number of diverse and increasingly broader levels or types of emotional response. To begin with, there is the startle reaction. Second, the dumb astonishment or stupefied amazement discussed at length by Charles Darwin in *The Expressions of the Emotions in Man and Animals* (chap. 12). Third, the various forms of goal-oriented curiosity and exploratory behavior. And last, the wondering attitude *stricto sensu*, which may seamlessly shade off, through awe, into nonpossessive love, doubt, puzzlement or sheer terror.

Let us begin with the startle reaction. Following the persuasive analysis by Jenefer Robinson,[33] I propose to understand it not as a bare and "blind" reflex—such as the proverbial knee-jerk reflex—but as an oriented and meaning-laden elementary response to our intentional environment. As such, it can be placed at the primitive end of a spectrum where at the opposite end lies the wondering mood.[34] The similarities are plain. Both kinds of feeling, respectively, alert us or attune us to a situation, the first one in an automatic, punctual, and invariant way, the other in a holistic and more articulated one. According to Robinson, "startlement" "is the way in which the organism focuses on a sudden loud sound and registers it as a significant piece of information: the response 'tells' the organism that the event is a significant event, something potentially dangerous which needs investigating further."[35] We might, therefore, imagine it as a bodily registration of significance that orients and articulates (though in nonlinguistic terms) the field of action of an animal being, in a manner not different from other emotional responses, given that

> an emotional response [is] a response that focuses our attention on (makes salient) and registers as significant to the goals (wants, motives) of the organism something in the perceived (remembered, imagined) environment; this response characteristically consists in motor and autonomic nervous system changes, including changes in the facial musculature, which may be part of a motivated action sequence (such as fight or flight) or function chiefly as communication to oneself and others of the state of the organism. In humans (and perhaps other species) the response is sometimes subjectively experienced as a recognizable "feeling."[36]

Even though I do not sympathize with Robinson's basically dualist and reductivist bias, it is indeed noteworthy that the bodies of humans (and those of a number of other animals) automatically work as a kind of warning radar or sensor capable of detecting surprising facets or features of the environment. This seems to corroborate the idea that we are creatures in whom the wondering attitude is deeply (both physically and evolutionarily) "ingrained." The fact that, sometimes, we involuntarily start at the world's sudden and unforeseen happenings prepares the ground for more articulated and sophisticated wonder-reactions. That is probably the reason why Charles Darwin went so far as to write in one of his unpublished notebooks that he could not "help thinking horses *admire* a wide prospect."[37] In fact, it is precisely the changing *quality* of their attention that seems to justify such a (at least at first sight) baffling hypothesis. If an animal can be "startled" by a surprising event, how much does it take for him to be able to respond with amazement to a sensory stimulus?

But the startle response is only one facet of the wondering attitude in its largest sense, and the least articulated one. Akin to "startlement," but a little more elaborate, is the reaction of surprise described by Darwin as a sudden increase of attention in *The Expression of the Emotions* (chap. 12):

> Attention, if sudden and close, graduates into surprise; and this into stupefied amazement. The latter frame of mind is closely akin to terror. Attention is shown by the eyebrows being slightly raised; and as this state increases into surprise, they are raised to a much greater extent, with the eyes and mouth widely opened. . . . As surprise is excited by something unexpected or unknown, we naturally desire, when startled, to perceive the cause as quickly as possible; and we consequently open our eyes fully, so that the field of vision may be increased, and the eyeballs moved easily in any direction.[38]

If the difference between "startlement" and astonishment may be described as a different "style of partiality,"[39] namely, as an embodied growth in world-openness (startlement being more defensive, and stupefied amazement more inquisitive), then on the next step of the scale can be placed curiosity, because of its aimless and voracious nature. Traditionally, curiosity has been regarded as a questionable and suspect frame of mind, wavering between virtue and vice.[40] Augustine famously depicted it in the *Confessions* as a "variety of lust," "akin to bodily incontinence" and prompted by pride.[41] Several centuries later, Martin Heidegger revived the topos by projecting it against the background of his controversial contrast between authentic and inauthentic ways of being, when he wrote that curiosity "seeks novelty only in order to leap from it anew to another novelty . . . abandoning itself to the world. . . . In not tarrying, curiosity is concerned with the constant possibility of distraction. Curiosity has nothing to do with observing entities and marvelling at them."[42] But, inasmuch as the distinction between authentic and inauthentic is renowned for its unreliability, it is probably more useful to leave aside any normative judgment and lay emphasis on the crucial role that gratuitous curiosity plays in animal life in general.

Having been portrayed for centuries as a light, playful, impudent, restless state of mind and the source of an often indiscriminately explorative behavior (curiosity was still for Burke "the first and simplest emotion which we discover in the human mind" and "the most superficial of all the affections"),[43] curiosity underwent a characteristic (albeit contested) early modern reevaluation, beginning with Hobbes, who in the *Leviathan* (chap. 6) represented it as the quality that best distinguishes man from beast and as an inexhaustible source of delight. Curiosity's main fault has been traditionally imputed to its fickleness, inconsistency, whirling dynamism. Being, as Augustine stated, a form of *concupiscentia oculorum*, it appeared as constantly vulnerable to distraction and having the deplorable tendency of "never dwelling anywhere."[44] (A contemporary psychiatrist would likely diagnose it as a form of attention deficit hyperactivity disorder!) What is interesting is that, as acutely noted by Lorraine Daston and Katherine Park in their pioneering work *Wonders and the Order of Nature*, to a modern eye this voracious and obsessive lust of the senses is bound to look more akin to greed than to sexual desire, since "unlike lust, which aims at satisfaction, avidity aims at perpetuation of desire, darting from object to object, barely pausing to enjoy any of them."[45] Seen from this angle, human curiosity appears, then, to yield to the appeal of the world in a peculiar way: it hankers after totality, but it sticks to a dynamic and sequential regime, having no space for the out-of-time fullness of the *nunc stans*. Curiosity, therefore, embodies in its own way a passion for totality, but only insofar as it works as a pure conatus, an unquenchable desire or "a highly refined form of consumerism."[46]

Thus relying on a simple contrast between the ancients' contempt (curiosity as "the beginning of all sin") and the moderns' commendation ("continuall and indefatigable generation of knowledge," to quote Hobbes again) is very far from telling the true story. Idle curiosity, as the old saying goes, is likely to kill the cat, but in its theoretical guise it also paved the way for the Scientific Revolution.[47] Curiosity has a characteristic Janus face, and, possibly, the young Darwin had something of this in mind when he amusingly pondered about the "scientific" nature of a foxy avatar of the proverbial cat on the island of San Pedro (Chile):

> In the evening we reached the island of San Pedro, where we found the Beagle at anchor. In doubling the point, two of the officers landed to take a round of angles with the theodolite. A fox (Canis fulvipes), of a kind said to be peculiar to the island, and very rare in it, and which is a new species, was sitting on the rocks. He was so intently absorbed in watching the work of the officers, that I was able, by quietly walking up behind, to knock him on the head with my geological hammer. This fox, *more curious or more scientific*, but less wise, than the generality of his brethren, is now mounted in the museum of the Zoological Society.[48]

The main difference between curiosity and wonder lies exactly in the quieter and more contemplative attitude of the second one. Wonder is an exceptionally intense way of being affectively aware of the things that surround us. In this sense, it results in a powerful experience of presence: presence of the object (and its horizon of meaning) and presence to herself of the subject (as a form of self-unity, not of self-consciousness), made possible by the unexpected emergence of a nondissipative time: a dilated present. The upshot is a combination of activity and passivity, which not only reflects the paradoxical fact that, in Cato's words (which Arendt was fond of quoting), "never is one more active than when one does nothing" (*numquam se plus agere quam nihil cum ageret*), but is contingent upon the world's being, as it were, the reason (or the occasion) for and not the efficient cause of the wondering response. In other words, in wondering the subject is absorbed by reality, without being its hostage or puppet. In this sense, the wonder-response always embodies a form of assent, a "yea-saying," and, having no utility whatsoever, it fosters in the subject a vague sense of joy in the very fact of being alive and of an objectless gratitude that turns outward. Wonder is an expansive response to the world's allure that encourages respect, compassion, gentleness, humility, unpossessiveness. As Jeff Malpas aptly put it, "The experience of wonder is an experience of our being already *given over to* the world and the things in it."[49]

The aforementioned blend of activity and passivity is also the result of another intriguing aspect of the wonder experience: its recursiveness. While the wonder-response is always originated by and focused on a specific object, it rapidly elicits a potentially infinite play of resonances. Indeed, we react with astonishment to a whole web of relations of likeness and difference that hints at the world's superabundance of meaning. This is, as it were, the background or resonating wonder that operates as an expanding halo around any particular source of amazement.[50] This is the kind of "oceanic" feeling alluded to in the previous discussion of moods.

As already hinted, this All-feeling must be seen as a very ordinary experience, since it is usually triggered by something that is not extraordinary or unusual at all. The idea is magnificently expressed by Czeslaw Milosz in the first lines of one of his most personal books:

> I am here. Those three words contain all that can be said—you begin with those words and you return to them. Here means on this earth, on this continent, in this city and no other, and in this epoch I call mine, this century, this year. I was given no other place, no other time, and I touch my desk to defend myself against the feeling that my own body is transient. This is all very fundamental, but, after all, the science of life depends on the gradual discovery of fundamental truths.
>
> I have written on various subjects, and not, for the most part, as I would have wished. Nor will I realize my long-standing intention this time. But I am always aware that what I want is impossible to achieve. I would need the ability to communicate my full amazement at "being here" in one unattainable sentence which would simultaneously transmit the smell and texture of my skin, everything stored in my memory, and all I now assent to, dissent from.[51]

Of course, when we are amazed by the simple fact of "being here," we react to an entire gamut of associations that verges on the infinite. We could also say that in these kinds of experiences we are faced with a special resistance of reality to thought, which forces us to practice a new and unsettling economy of attention. In a recent essay, Cora Diamond has dubbed this clash against the limits of philosophical discourse "the difficulty of reality."[52] The most disturbing aspect of this experience is the concurrence of ordinariness and exceptionality. Someone can be struck dumb by the simple fact of the existence and doom of other animals ("beings so like us, so unlike us, so astonishingly capable of being companions of ours and so unfathomably distant")[53] or by the simultaneous presence of life and death in our ordinary lives, while, for others, such things fail even to raise an eyebrow. Who is right and who is wrong? Where is the boundary to be drawn between the ordinary and the exceptional in our lives? The feeling of wonder is our main way of coping with such a distressing situation. When we yield to wonder, we stop and let ourselves be absorbed for a while into the world's complexity. And this kind of capitulation is the source of a broken knowledge: a botched understanding of reality, which is yet respectful of its contradictory richness, although one can go mad trying "to bring together what cannot be thought."[54]

It is crucial not to lose sight of the instability and the general destabilizing effect of the whole gamut of wonder-responses. The startle response is, of course, a bodily destabilizing experience (this is actually its main biological function). And the same can be said of stupefied amazement, with its widely open eyes and mouth, the full and deep inspiration, the uplifting of the hands, and so forth. Curiosity, in its turn, is unsettling in the sense in which Hegel's "bad infinite" is unsettling: it spawns an endless series moving from one thing to another, like a straight line with no end. And wonder is an unstable emotion because it is always on the verge of turning into other fundamental moods such as joy (elation or serenity), anxiety (objectless fear), melancholy, and boredom (indifference).[55] Indeed, reality's superabundance of meaning, which may stir up admiration or astonishment, may as easily bring about a sense of euphoria, uneasiness (if not sheer horror), calm sadness, or nausea.[56] Wonder cannot last for long, and this should say something about the human condition. But what?

## Comic Enchantment

Usually, when one thinks about the disenchantment or reenchantment of the world, global images loom on the horizon, as if Weber's *Zeitdiagnose* were a matter of all or nothing: magic or science, belief or unbelief, poetry or prose, courage or cowardice, "intellectual sacrifice" or "intellectual integrity." But, *ça va sans dire*, our lives are a great deal richer than that and could hardly be contained in such a narrow picture of reality.

There is some evident truth in Weber's historical diagnosis—albeit its true interpretation remains uncertain—but it is now also clear that, even in highly rationalized modern societies, forms and moments of enchantment survive (and sometimes thrive) within finite provinces of meaning, to use Alfred Schutz's memorable phrase.[57] In the following and concluding pages, I will not talk about the most obvious candidates for modern enchantment—art and technology—but I will dwell upon a very ordinary and pervasive kind of wonder-experience: humor.[58]

Even apart from the significant semantic link between humor as a mental disposition (or state of mind) and humor as the faculty of perceiving what is ludicrous or amusing in things, it should be easy to discern the "moody" quality of wit—understood as a distinctive attitude toward reality—and its affinity with wonder. After all, everybody knows how it feels to be in a hilarious mood, and how that alters our view of reality. The sense of humor discloses a hidden world; it looks beyond the obvious and, in doing so, temporarily transfigures our ordinary perception of things. Laughter is a special way of responding to the world and of seeing it *as* something new and different from the world of our daily life: as funny, odd, incongruous.[59] Something that is likely to "shoulder out one's body" or, better, make one gently collapse into one's own body, to use Helmuth Plessner's favorite image.

This incongruity effect is actually the product of an overlapping of pictures or, in Arthur Koestler's words, "bisituation," that is to say "the capacity to associate, to draw together two (or more) previously nonassociated aspects of reality."[60] The incongruity produces a double gaze, partially detached or interestedly contemplative, that leads to an ocscillation around the center that is one's own being, and thus to a feeling of giddiness. According to Kant,

> laughter is an affect that arises if a tense expectation is transformed into nothing . . . when the illusion vanishes into nothing, the mind looks into the illusion once more in order to give it another try, and so by a rapid succession of tension and relaxation the mind is bounced back and forth and made to sway; . . . then we can pretty well grasp how, as the mind suddenly shifts alternately from one position to another in order to contemplate its object, there might be a corresponding alternating tension and relaxation of the elastic part of our intestines that is communicated to the diaphragm.[61]

Kant's amusing and delightful explanation of laughter captures well the special nature of the comic catharsis and its proximity to the "difficulty of reality." The bodily response of laughter provoked by the impact of some whirling bi-stable image is one of the most effective ways of coping with the multidimensionality of our world experience. It is an attempt to think beyond the (affective and cognitive) limits of our finite animal condition, which, however, leads to "nothing" (of positive) or, better, to a sort of "broken knowledge," a second-order understanding that amounts to a net gain in reflexivity. For, with a comic insight, what changes is not a single mental representation, but our whole epistemic background, our stance toward or, better, attunement with the world. We might also speak of an ordinary transfiguration of reality, which continues to be the same, though it now looks different and, in a sense, more enigmatic. After a good laugh, we feel more at home in the world without necessarily being reconciled with it.[62] As a matter of fact, the comic catharsis does not solve any of our existential riddles, but it helps us to come to terms with our strained ex-centrical nature, even though our sense of humor grants us only ambivalent answers and leaves us with a feeling of uncertainty and vulnerability.

Of course, a world inhabited only by buffered selves (i.e., individuals with stable, sturdy, impervious identities) would basically be a world shorn of humor and wonder. But the truth is that even our disenchanted modern world continues to be a world full of both, and this is mainly due to its fractured (and often incongruous) horizons. As noted by Peter Berger: "Modernity *pluralizes* the world. It throws together people with different values and worldviews; it undermines taken-for-granted traditions; it accelerates all processes of change. This brings about a multiplicity of incongruencies—and it is the perception of incongruence that is at the core of the comic experience."[63]

In short, the comic enchantment may be understood as a peculiarly "secular" and impertinent variety of wonder (the direct opposite of awe) in two different senses. First, because it embodies an ordinary, "low," bodily, impure, sensual, concerned, unruly, fickle form of contemplative detachment: an empowering and "inverted sublime," to use Jean Paul's ingenious and often-quoted formula. Indeed, its lack of emphasis and of tragedy, its unsteadiness and elusiveness, mirror a crucial aspect of the human condition as portrayed above, that is, the pluralism, plasticity, and perspectivism inherent in the incessant dialectic of the moods that every human being has to deal with. In this sense, through our sense of humor we are attuned to the "core" of reality, as it is disclosed to our sensitivity. Second, we are entitled to speak of a "secular" wonder because it is a form of enchantment especially suited for a modern "secular" age; which does not necessarily mean an unbelieving or unspiritual time, but one in which, nonetheless, the general tone is set by a fractured, restless, horizontal, skeptical, urbane, and ironic climate.[64]

So how does this leave us with respect to our opening question: what is it like to be a secular creature?

Our journey through the human emotional landscape has taken us a long way from the tragic (and ultimately self-defeating) undertone of Silenus's reply. Not so far, however, that we can fail to recognize its partial truth. We humans are not blessed creatures—needy and fragile as we are—but neither are we powerless or, least of all, cursed. Our predicament seems rather to fall under the sign of unsteadiness, ambivalence, elusiveness, or, to employ a now-familiar word, incongruity. Neither blessing, nor curse, but a sort of unending trial or challenge. That is why, all things considered, the injunction to choose between the usual anti-thetical options—enchantment or disenchantment, adult endurance or childish helplessness, secularism or religion, embeddedness or homeless-ness—does not sound convincing in the least. In the world's superabun-dance of meaning, it is the Gestalt shift of a tri-stable or, better, multi-stable image that seems to give the general tone to our lives. A constant swinging between wonder, joy, indifference, melancholy, and fear rules our lives and generally leaves us *depaysé* and baffled; and in need (at least temporarily) of a balancing point. As a matter of fact, because of the constitutive role played by the world in our way of being, we cannot but struggle to feel at home in it, even though such an abode has many facades and some are not reassuring at all.[65] And that is why most of us are, most of the time, restless and dissatisfied. We habitually feel at home only *by* dreaming of being elsewhere. And that looks like a laughing mat-ter, in its own way.

In this scenario, it may be helpful, if not vital, to rely on a "comic" anthropology, that is, on an anthropology of ambiguity that recoils from both optimistic and pessimistic pictures of man. Laughter, like wonder, is not bound to last, but its mixed, median, and fleeting nature seemingly represents an appropriate response to the eccentric and indeterminate ontological status of human beings.

One might speculate, therefore, whether we should not seek the type of outlook suitable for grasping the point of human nature, the image that defines its horizon of intelligibility, in comedy before looking to tragedy or the tragic range. In our Western tradition, the greatest interest seems to lie in tragic events that affect human beings by chance, and that have been splendidly evoked in the dramas from all periods as well as in the masterpieces of the modern novel. Such an interest is doubtlessly justified by the important role moral dilemmas play in ethical reflection. If you consider it carefully, however, the life of individuals is not made up only of tragic dilemmas. It is often dotted with curious indecisions, funny mis-haps, or situations marked by a comic turn of events completely interwo-ven with its most melancholic aspects. You could hazard the hypothesis that, in many cases, a solution to the dilemmas of existence is hidden in the ability to grasp the subtly comic aspect of many human events. In-deed, there is something essentially alive in a laugh, and, without doubt, it hides an important answer to the awkward condition of every human being, an alternative to the inflexibility of shame and disgust.

To this end, there is a long tradition of comic works from Aristo-phanes, Plautus, Terence, through Rabelais, Molière, Cervantes, Fielding,

Smollett, and in our day Milan Kundera or Jonathan Coe. All these works are dominated by a taste for and an almost unrestrained curiosity about human nature. It is not just by chance that in the first chapter of *Tom Jones*, Fielding uses the luxuriant sensual image of a banquet to attract the reader to approach his work. All the same, the menu provides for a single dish: human nature, but in it "is such prodigious variety that a cook will have sooner gone through all the several species of animal and vegetable food in the world than an author will be able to exhaust so extensive a subject." Indeed, the image of a symposium, with its confusion, carnality, capricious regularity lends itself wonderfully to illustrating our mixed, impure nature, foreseeable even though it cannot be totally objectified.

The comic character of the human condition, however, need not necessarily be detrimental to man's dignity. The contrast between the two is really based on a misunderstanding and a narrow interpretation of the human ability to look at one's own vicissitudes with delight and irony. After all, there is also a comic way to interpret Pico della Mirandola's *Oratio de hominis dignitate* and his image of a man without a fixed place in the world, tossed up and down the steps of the ladder of being. You can even imagine that the face of the *optimus opifex* wears a smile when he addresses man with the famous sentences

> Neither a fixed abode nor a form that is thine alone nor any function peculiar to thyself have we given thee, Adam, to the end that according to thy longing and according to thy judgment thou mayest have and possess what abode, what form, what functions thou thyself shalt desire. The nature of all other beings is limited and constrained within the bounds of laws prescribed by Us. Thou, constrained by no limits, in accordance with thine own free will, in whose hand We have placed thee, shalt ordain for thyself the limits of thy nature. We have set thee at the world's center that thou mayest from thence more easily observe whatever is in the world. We have made thee neither of heaven nor of earth, neither mortal nor immortal, so that with freedom of choice and with honor, as though the maker and molder of thyself, thou mayest fashion thyself in whatever shape thou shalt prefer. Thou shalt have the power to degenerate into the lower forms of life, which are brutish. Thou shalt have the power, out of thy soul's judgment, to be reborn with the higher forms, which are divine.[66]

After all, in the being depicted by Pico, *indiscretae opus imaginis*, you could—why not?—even make out the familiar characteristics of Charlie Chaplin, the enterprising schlemiel "who does not acknowledge the hierarchical structures of the world,"[67] and in whose clumsiness, intelligence, ill luck, and craftiness anyone can recognize him- or herself with a smile, and learn to observe with participation, admiration, and indulgence human efforts to transcend the needy condition of human beings at grips with a world that is at times hostile, at times lavish, and which, although it often thwarts our excessive ambitions, sometimes can satisfy us beyond our wildest dreams.

# 7

## ❧

# Prehuman Foundations of Morality

### *Frans B. M. de Waal*

If we look at our species without letting ourselves be blinded by the technical advances of the last few millennia, we see a creature of flesh and blood with a brain that, albeit three times larger than a chimpanzee's, doesn't contain any new parts. Superior our intellect may be, but we have no basic wants or needs that cannot also be observed in our close relatives. I interact on a daily basis with apes, such as chimpanzees and bonobos, which are known as *anthropoids* precisely because of their humanlike characteristics. Like us, they strive for power, enjoy sex, want security and affection, kill over territory, and value trust and cooperation. Yes, we use cell phones and fly airplanes, but our psychological makeup remains that of a social primate.

But does this also include our prized morality? Human morality is considered so special that geneticist Francis Collins has taken its presence as proof that God exists, thus following in the footsteps of C. S. Lewis.[1] No wonder, claiming morality as a "mere" product of biology is seen as a threat to religion as well as to morality itself, as if such a view would absolve us from the obligation to lead virtuous lives. In a recent example, an American reverend, Al Sharpton, opined: "If there is no order to the universe, and therefore some being, some force that ordered it, then who determines what is right or wrong? There is nothing immoral if there's nothing in charge."[2] Similarly, I have heard people echo Dostoevsky's Ivan Karamazov, exclaiming that "[i]f there is no God, I am free to rape my neighbor!"

Perhaps it is just me, but I am wary of anyone whose belief system is the only thing standing between them and repulsive behavior. Why not assume that our humanity, including the self-control needed to form livable societies, is built into us? Does anyone truly believe that our ancestors lacked rules of right and wrong before they had religion? Did they never assist others in need, or complain about an unfair deal? Human morality must be quite a bit older than religion and civilization. It may, in fact, be older than humanity itself. Other primates live in highly structured social groups in which rules and inhibitions apply and mutual aid is a daily occurrence. Acts of genuine kindness do occur in animals as they do in humans, because once a tendency has been put into place by nature, it is not essential that each and every expression of it serve survival and reproduction. It is a bit as with the sex drive: it evolved to serve reproduction, but that does not mean that humans and animals have sex only in order to reproduce. The behavior follows its own autonomous motivational dynamic. Arguments such as Collins's often rest on a conflation of the reasons why a behavior evolved and the reasons why individual actors show it, a distinction as sacred to biologists as the one between church and state in modern society. In the same way, altruistic behavior has been put into place to serve a cooperative group life, which benefits the actors of such behavior, yet in daily life follows its own autonomous motivations, which vary from self-serving to other-regarding.[3]

Without claiming other primates as moral beings, it is not hard to recognize the pillars of morality in their behavior. These pillars are summed up in our golden rule, which transcends the world's cultures and religions. "Do unto others as you would have them do to you" brings together empathy (attention to another's feelings) and reciprocity (if others follow the same rule, you will be treated well). Human morality could not exist without empathy and reciprocity, tendencies that are recognizable in our fellow primates.

## Animal Empathy

### Emotional Linkage

When Zahn-Waxler visited homes to find out how children respond to family members instructed to feign sadness (sobbing), pain (crying), or distress (choking), she discovered that children a little over one year of age already comfort others. This is a milestone in their development: an aversive experience in another person draws out a concerned response. An unplanned sidebar to this study, however, was that some household pets appeared as worried as the children by the "distress" of a family member. They hovered over them or put their heads in their laps.[4]

When the emotional state of one individual induces a matching or related state in another, we speak of *emotional contagion*.[5] With increasing differentiation between self and other, and an increasing appreciation of the precise circumstances underlying the emotional states of others, emotional contagion develops into empathy. Empathy encompasses—and could not possibly exist without—emotional contagion, yet it goes further in that it places filters between the other's state and one's own, and adds a cognitive layer. In empathy, the subject realizes that its own internal state may be different from the other's.[6]

Empathy is a social phenomenon with great adaptive significance for animals in groups. That most modern textbooks on animal cognition do not index empathy or sympathy does not mean that these capacities are not essential; it only means that they have been overlooked by a science traditionally focused on individual rather than interindividual capacities. Inasmuch as the survival of many animals depends on concerted action, mutual aid, and information transfer, selection must have favored proximate mechanisms to evaluate the emotional states of others and quickly respond to them in adaptive ways. Even though the human empathy literature often emphasizes the cognitive side of this ability, Hoffman rightly noted that "humans must be equipped biologically to function effectively in many social situations without undue reliance on cognitive processes."[7]

Empathy allows us to relate to the emotional states of others. This is critical for the regulation of social interactions, such as coordinated activity, cooperation toward a common goal, social bonding, and care of others. It would be strange indeed if a survival mechanism that arises so early in life in all members of our species were without animal parallels.

Preston and de Waal propose that at the core of the empathic capacity is a relatively simple mechanism that provides an observer (the "subject") with access to the subjective state of another (the "object") through the subject's own neural and bodily representations.[8] When the subject attends to the object's state, the subject's neural representations of similar states are automatically activated. The more similar subject and object, the more the object will activate matching peripheral motor and autonomic responses in the subject (e.g., changes in heart rate, skin conductance, facial expression, body posture). This activation allows the subject to understand that the object also has an extended consciousness including thoughts, feelings, and needs, which in turn fosters sympathy, compassion, and helping.

## *Evidence of Primate Empathy*

Qualitative accounts of great apes support the view that these animals show strong emotional reactions to others in pain or need. Yerkes reported how his bonobo, Prince Chim, was so extraordinarily concerned and protective toward his sickly chimpanzee companion, Panzee, that the scientific establishment might not accept his claims: "If I were to tell of his altruistic and obviously sympathetic behavior towards Panzee I should be suspected of idealizing an ape."[9] Ladygina-Kohts noticed similar empathic tendencies in her young chimpanzee, Joni, whom she raised at the beginning of the previous century, in Moscow.[10] Kohts, who analyzed Joni's behavior in the minutest detail, discovered that the only way to get him off the roof of her house after an escape (much better than any reward or threat of punishment) was by appealing to his sympathy:

> If I pretend to be crying, close my eyes and weep, Joni immediately stops his play or any other activities, quickly runs over to me, all excited and shagged, from the most remote places in the house, such as the roof or the ceiling of his cage, from where I could not drive him down despite my persistent calls and entreaties. He hastily runs around me, as if looking for the offender; looking at my face, he tenderly takes my chin in his palm, lightly touches my face with his finger, as though trying to understand what is happening, and turns around, clenching his toes into firm fists.

I have discussed similar reports, suggesting that apart from emotional connectedness apes have an appreciation of the other's situation and a degree of perspective taking.[11] O'Connell conducted a content analysis of thousands of qualitative reports, counting the frequency of three types of empathy, from emotional contagion to more cognitive forms, including an appreciation of the other's situation.[12] Understanding the emotional state of another was particularly common in the chimpanzee, with most outcomes resulting in the subject's comforting the object of distress. Monkey displays of empathy were far more restricted, but did include the adoption of orphans and reactions to illness, handicaps, and wounding.

This difference between monkey and ape empathy is also evident from studies of behavior known as "consolation," first documented by de Waal and van Roosmalen.[13] Consolation is defined as friendly, reassuring contact directed by an uninvolved bystander at one of the combatants in a previous aggressive incident. For example, a third party goes over to the loser of a fight and gently puts an arm around his shoulders (fig. 1).

Fig. 1. Consolation among chimpanzees: a juvenile puts an arm around a screaming adult male who has just been defeated in a fight with a rival. Consolation is common in apes, but rare or absent in the monkeys. Photograph by the author.

Consolation is not to be confused with reconciliation, which seems selfishly motivated (see above). The advantages of consolation for the actor remain unclear. The actor could probably walk away from the scene without negative consequences.

Chimpanzee consolation is well quantified. De Waal and van Roosmalen based their conclusions on an analysis of hundreds of postconflict observations, and a replication by de Waal and Aureli included an even larger sample in which the authors sought to test two relatively simple predictions.[14] If third-party contacts indeed serve to alleviate the distress of conflict participants, these contacts should be directed more at recipients of aggression than at aggressors, and more at recipients of intense than of mild aggression. Comparing third-party contact rates with baseline levels, we found support for both predictions. A recent study by Fraser et al. confirmed that consolation behavior reduces the stress of its recipients.[15]

## Prosocial Tendencies

An aging female, named Peony, spends her days with other chimpanzees in a large outdoor enclosure near Atlanta, Georgia. On bad days, when her arthritis is acting up, she has great trouble walking and climbing. But other females help her out. For example, Peony is huffing and puffing to get up into the climbing frame in which several chimpanzees have gathered for a grooming session. An unrelated younger female moves behind her, places both hands on her ample behind, and pushes her up with quite a bit of effort, until Peony joins the rest.

Even though examples of spontaneous helping among primates are abundant (see above), this is not the impression one gets from the modern literature according to which humans are the only truly altruistic species, since all that animals care about are return-benefits.[16] The evolutionary reasons for altruistic behavior are not necessarily the animals' reasons, however. Do animals really help each other in the knowledge that this will ultimately benefit themselves? To assume so is cognitively demanding in the extreme, requiring animals to have expectations about the future behavior of others and to keep track of what they did for others versus what others did for them. Thus far, there is little or no evidence for such expectations. Helpful acts for immediate self-gain are indeed common, but it seems safe to assume that future return-benefits remain largely beyond the animal's cognitive horizon.[17]

The motivation to help must stem from immediate factors, such as a sensitivity to the emotions and/or needs of others. Such sensitivity would by no means contradict self-serving reasons for the evolution of behavior so long as it steers altruistic behavior into the direction predicted by theories of kin selection and reciprocal altruism. In humans, the most commonly assumed motivation behind altruism is empathy. We identify with another in need, pain, or distress, which induces emotional arousal that may translate into sympathy and helping.[18] Inasmuch as there are ample signs of empathy in other animals, the same hypothesis may apply. This can be tested through evaluation of how animals perceive another's situation, and under which circumstances they try to ameliorate this situation.

Apart from assisting an aging female in her climbing efforts, chimpanzees occasionally perform extremely costly helping actions. For example, when a female reacts to the screams of her closest associate by defending her against a dominant male, she takes enormous risks on behalf of the other. She may very well get injured. Note the following description of two longtime chimpanzee friends in a zoo colony: "Not only do they often act together against attackers, they also seek comfort and reassurance from each other. When one of them has been involved in a painful conflict, she goes to the other to be embraced. They then literally scream in each other's arms."[19] This kind of cooperation, expressed in alliances and coalitions, is among the best documented in primatology.[20]

Recent studies have set out to determine the precise circumstances under which primates are willing to assist either humans or each other. The investigators tried to rule out reciprocity by having the apes interact with humans they barely knew, and on whom they had never depended for food or other favors. They also tried to rule out the role of immediate return-benefits by manipulating the availability of rewards. The chimpanzees spontaneously assisted persons regardless of whether or not this yielded a reward, and were also willing to help fellow chimpanzees reach a room with food. One would think that rewards, even if not strictly necessary, would at least stimulate helping behavior, but in fact they seemed to play no role at all. Since the decision to help did not seem to be based on a cost/benefit calculation, it may have been genuinely other-oriented.[21]

A recent study further demonstrated spontaneous helping in marmosets,[22] and we ourselves conducted a similar experiment with capuchin monkeys. We placed two monkeys side by side: separate, but in full view. One of them needed to barter with us with small plastic tokens, which we'd first give to a monkey, after which we'd hold out an open hand, letting them return the token for a tidbit. The critical test came when we offered a choice between two differently colored tokens with different meaning: one token was "selfish," the other "prosocial." If the bartering monkey picked the selfish token, it received a small piece of apple for returning it, but its partner got nothing. The prosocial token, on the other hand, rewarded both monkeys equally at the same time (fig. 2). Since the monkey who did the bartering was rewarded either way, the only difference was in what the partner received. To make sure they understood, we would make quite a show by either raising one hand with food and feeding one monkey, or raising both hands and simultaneously handing food to both of them.

We know exactly how socially close any two monkeys are because we watch how much time they spend together in the group. We found that the stronger the tie with its partner, the more a monkey would pick the prosocial token. The procedures were repeated many times with different combinations of monkeys and different sets of tokens, and they kept doing it. Their choices could not be explained by fear of punishment, because in every pair the dominant monkey (the one who had least to fear) proved the more prosocial one.[23]

In short, there is mounting evidence that primates do care about each other's welfare, and do take actions to improve it, probably based on empathy with the other and the self-rewarding qualities of helping behavior. Since empathy derives from state matching, it automatically produces a stake in the other's welfare, which may explain the "warm-glow" effect, that is, pleasant feelings associated with improvement of another's condition. When humans do good, they report good feelings, and show activation of reward-related brain areas,[24] and it will be important to determine whether the same self-reward system operates in other primates.

## Reciprocity

Chimpanzees and capuchin monkeys—the two species I work with the most—are special, because they are among the very few primates that share food outside the mother-offspring context. The capuchin is a small animal, easy to work with, as opposed to the chimpanzee, which is many times stronger than we are. But chimpanzees, too, are interested in each other's food and will share on occasion—sometimes even hand over a piece of food to another. Most sharing, however, is passive, where one individual will reach for food owned by another, who will let go. But even such passive sharing is special compared to the behavior of most animals, for whom such a situation might result in a fight or assertion by the dominant, without sharing.

One series of experiments concerned the idea that monkeys cooperate on the basis of mental record keeping of favors. We set up a situation to study tit for tat: I do something for you and, a while later, you do something for me. Inspired by a classic 1930s study at the Yerkes Primate Center and the theories of Trivers,[25] we presented a pair of capuchin monkeys with a tray with two pull bars attached to it. Both monkeys sat in a test chamber with mesh between them, so that they could see each other and share food through the mesh. The tray was counterweighted such that a

Fig. 2. One capuchin monkey reaches through an armhole to choose between differently marked pieces of pipe while another looks on. The pipe pieces can be exchanged for food. One token feeds both monkeys; the other feeds only the chooser. Capuchins typically prefer the "prosocial" token. Drawing by Frans de Waal.

single monkey couldn't pull it: they needed to work together. Only one side was baited, meaning that only one of the two monkeys would obtain a food reward.

After successful pulls we measured how much food the possessor shared with its helper. Possessors could easily keep the food by sitting in the corner and eating alone, but didn't do so. We found that food sharing after cooperative efforts was higher than after solo efforts. That is, the possessor of food shared more with the monkey on the other side of the mesh if this partner had played a role in securing the food than if the possessor had acquired the food on its own. Capuchins thus seem to reward helpers for their efforts, which is of course also a way of keeping assistants motivated.[26]

Investigators can also explore reciprocity by handing one chimpanzee a large amount of food, such as a watermelon or leafy branch, and then observing what follows (fig. 3). As in the 1980s hit song "What Have

You Done for Me Lately?" the apes seem to recall previous favors, such as grooming. We analyzed no fewer than seven thousand approaches to food owners to see which ones met with success. During the mornings before every feeding session, we had recorded spontaneous grooming. We then compared the flow of both "currencies": food and grooming. If the top male, Socko, had groomed May, for example, his chances of getting a few branches from her in the afternoon were greatly improved. We found this effect all over the colony: apparently, one good turn deserves another. This kind of exchange must rest on memory of previous events combined with a psychological mechanism known as "gratitude"; that is, warm feelings toward someone whose act of kindness we recall. Trivers had already speculated about this mechanism, which seems well developed in chimpanzees.[27]

## Fairness

The above relates to the distribution of payoffs. How skewed can it be before cooperation disappears? According to a recent theory, the well-known human aversion to inequity relates to the need to maintain cooperation.[28] Similarly, cooperative nonhuman species seem guided by a set of expectations about payoff distribution. I have proposed a *sense of social regularity*, defined as "[a] set of expectations about the way in which oneself (or others) should be treated and how resources should be divided. Whenever reality deviates from these expectations to one's (or the other's) disadvantage, a negative reaction ensues, most commonly protest by subordinate individuals and punishment by dominant individuals."[29]

We paired each monkey with a group mate and watched reactions if their partners got a better reward for doing the same bartering task. This consisted of an exchange in which the experimenter gave the subject a token that could immediately be returned for a reward. Each session consisted of twenty-five exchanges by each individual, and subjects always saw their partner's exchange immediately before their own. Food rewards varied from low-value rewards (a cucumber piece), which they are usually happy to work for, to high-value rewards (a grape), which were preferred by all individuals tested. All subjects underwent (a) an Equity Test, in which subject and partner did the same work for the same low-value food; (b) an Inequity Test, in which the partner received a superior reward (grape) for the same effort; (c) an Effort Control Test, designed to elucidate the role of effort, in which the partner received the higher-value

Fig. 3. A cluster of food-sharing chimpanzees at the Yerkes Field Station. The female in the top-right corner is the possessor. The female in the lower left corner is tentatively reaching out for the first time; whether or not she can feed will depend on the possessor's reaction. Photograph by Frans de Waal.

grape for free; and (d) a Food Control Test, designed to elucidate the effect of the presence of the reward on subject behavior, in which grapes were visible but not given to another capuchin.

Individuals who received lower-value rewards showed both passive negative reactions (i.e., refusal to exchange the token, ignoring the reward) and active negative reactions (i.e., throwing out the token or reward). Compared to tests in which both received identical rewards, in those in which their partner received a better deal the capuchins were far less willing to complete the exchange or accept the reward.[30] Capuchins refused to participate even more frequently if their partner did not have to work (exchange) to get the better reward but was handed it for "free."

Of course, there is always the possibility that subjects were reacting to the mere presence of the higher-value food, and that what the partner received (free or not) did not matter. In other words, they may simply be holding out for something better. To test this, we added a twist to our study. Before each Equity Test, in which both monkeys ate cucumber, we'd wave grapes around, just to show that we had them. This hardly bothered the monkeys: they still contentedly traded for cucumber. Only if the grapes were actually given to its partner did the one who missed out go into protest mode. It really was the inequity that bothered them.[31]

Capuchin monkeys thus seem to measure reward in relative terms, comparing their own rewards with those available and their own efforts with those of others. Given the reactions in 2009 to bonuses and salaries on Wall Street, the same sort of resentment is extremely well developed in humans, and probably forms the basis for our sense of fairness. This sense transcends our own personal interests, and as such seems more complex than what we observe in other primates, but it cannot be denied that sensitivity to the receipts of others compared to our own is how it probably all began.

## Conclusion

The idea that we would have needed a God, a creator, or another outside force that ordered the universe so as to arrive at a moral order is hard to maintain. It assumes that humans by themselves, left to their own devices, would have little interest in setting rules of conduct or promoting a modus vivendi. But humans evolved to live in groups, just as many other primates did, precisely because group life had survival value. We do not need an outside force to care about group life, therefore: we have plenty of interest of our own. Survival in groups requires a social life that promotes cooperation, protects the weak, and delivers benefits to all, which it does only if there are internal dynamics to ensure that conflicts are resolved and that everyone behaves in a matter consistent with group goals. This is precisely what morality does. Hence I see no reason to believe that religion has been essential in bringing this about. In other primates, we can see the beginning of a social motive, and humans have just taken it one step further.

What religion does, and does well, is take the prosocial tendencies of the human species, and its ability to reinforce and obey social norms, and bolster these tendencies by adding weight to them, mainly by providing the approval or disapproval of a higher being (among the many other theories about God by many important thinkers, Sigmund Freud and others have argued that God was modeled after a super alpha male, and speculated that when the father of the primal horde had been eliminated by his sons, he became the image of God).[32] Humans thus moved from a purely socially reinforced system to one with religious backing. A big step perhaps, but not big enough to claim morality as a religious invention.

Without claiming other primates as moral beings,[33] we may assert that the seeds for a moral order seem far older than our species. Empathy, sympathy, reciprocity, fairness, and other basic tendencies were built into humanity's moral order based on our primate psychology. We did not develop this order from scratch, but had a huge helping hand—not God's, but Mother Nature's.

# 8

## ❦

# The Truth Is Sacred

### *David Sloan Wilson*

The financial collapse of 2008 provided a vivid reminder that life as we know it depends upon a fabric of trust. When the fabric becomes frayed, we are in danger of being cast into an abyss more frightening than any fire-and-brimstone sermon—because this abyss is real.

As we peer into the abyss, consider the following quotation from Jim Cramer, TV personality and former hedge fund manager, whose show *Mad Money* is promoted by NBC with the slogan "In Cramer We Trust."

> What's important when you are in that hedge fund mode is to not be doing anything that is remotely truthful, because the truth is so against your view—it is important to create a new truth to develop a fiction ... you can't take any chances.

Cramer would never have said that on his television show; it comes from a 2006 interview that evidently put him in that hedge fund mode, blissfully unaware that the whole world might end up listening.

I was filled with righteous indignation when I first read this passage. I felt that Cramer had violated something sacred and deserved to be punished. I am a civilized man, so I didn't want to pull out his entrails while he was still alive (oops—that's what "civilized" Europeans did only a few centuries ago!), but I did want his punishment to be public and humiliating. I wanted him to be stripped of his social status, his influence, and any comforts that he might have received from his ill-gotten gains. Making him walk naked down Wall Street and Main Street seemed fitting. I would consider forgiveness only after he had proven himself to be truly repentant for his sins.

This volume of essays asks whether a purely secular view of the world can be as engaging and fulfilling as a religious worldview. The word "enchantment" has been used to describe what religion reliably delivers and what we hope secularism can also deliver. I have chosen to begin my essay by focusing on the negative pole of religion—wrath—rather than the positive pole of enchantment, but my core questions are the same. What is it about religion that sets it apart from secularism? What is worth wanting about religion, and can it be achieved by a belief system that fully respects factual reality?

I will begin by making a number of points that I regard as "Evolution 101." They are so elementary that they are unlikely to be false, yet they add up to conclusions that are far from obvious.

## Evolution 101

1. *The mind has evolved to adopt beliefs on the basis of practical, not factual, realism.* Any belief can be evaluated on the basis of two criteria: how well it corresponds to the real world (factual realism), and the degree to which it causes people to behave adaptively in the real world (practical realism). If the human mind evolved by natural selection, then we should create and adopt beliefs purely on the basis of *practical* realism. Factual realism is of no value whatsoever from an evolutionary perspective, except insofar as it leads to practical realism.

2. *The relationship between factual and practical realism can be positive or negative, depending upon the context.* In some situations, knowing the world as it really is can be highly adaptive. If I am hunting, I need to know the exact location of my prey. In other situations, beliefs can become *more* adaptive by *departing* from factual reality.[1] It is often more motivating to regard one's enemy as an inhuman monster than to view him as someone much like oneself. If the human mind is suitably adaptive, it should be able to *either* apprehend *or* depart from factual reality, depending upon the context. According to anthropologists such as Malinowski,[2] all cultures have a mode of thought that can be recognized as rational and protoscientific, which is expressed only some of the time. That is exactly what we should expect from an evolutionary perspective.

3. *Adaptive fictions include, but go beyond, beliefs in supernatural agents.* Emile Durkheim claimed that religious beliefs score high in practical realism, which he called *secular utility*.[3] No matter how bizarre and otherworldly, they help members of the religious community thrive in *this*

world. The modern study of religion from an evolutionary perspective has largely confirmed the Durkheimian view.[4] However, the gods are just the tip of the iceberg when it comes to adaptive fictions.

An example from the anthropological literature vividly makes this point. Micronesian islanders are master navigators, spending days out of sight of land as they travel between islands in their outrigger canoes. Unlike that of migratory birds, their ability to navigate is not genetically innate but based on a system of cultural beliefs, some of which appear crazy from the standpoint of factual realism. For example, they believe that the islands move while their boat stands still. They also believe in the existence of certain islands that do not, in fact, exist. Cognitive psychologist Edwin Hutchins has shown that these beliefs are highly adaptive, despite the fact that they are manifestly false. It is computationally more efficient for them to think of themselves as a single fixed point in a constellation of islands that move, than to think of themselves as a moving point in a constellation of fixed islands. And if there is not an actual island for them to imagine passing to keep track of their movement, they invent one.[5] This example is beautiful because it is so concrete. The superiority of the false beliefs can be demonstrated in computational terms, and we can easily imagine the factual realist, stubbornly insisting that the islands don't move and that certain islands don't exist, disappearing over the horizon in his canoe, never to return.

As a second example closer to home, Ayn Rand was one of the most influential intellectuals of the twentieth century, through her novels and her philosophy of objectivism. She called herself an atheist and rationalist, but her belief system can be shown to be structurally similar to fundamentalist religion, and she was proud to call it a *stylized universe*.[6] She made abundant use of adaptive fictions; the fact that she was not invoking supernatural agents was a mere detail.

To summarize this point, religion is not a dividing line when it comes to adaptive fictions; they are expected whenever there is a negative tradeoff between factual and practical realism.

4. *Adaptive fictions often, but not always, count as a form of selfishness.* So far I have been describing the consequences of beliefs for the individual believer, but they also have *social* consequences. All combinations are possible; a given belief might be good for the individual and bad for others, bad for the individual and good for others, good for both, or bad for both. When an adaptive fiction benefits everyone in a group or benefits individuals without harming others, there is little incentive to question it.

This is true not only for religious beliefs, but for the assumptions of many nonreligious belief systems.

Jim Cramer's comment is a clear case of a fiction that is good for the individual but bad for the group as a whole. Many such cases can be expected. Entire societies can collapse under the weight of such lies, which is why Cramer's blatant disregard for the truth is so frightening and maddening. When adaptive fictions count as a form of selfishness, we can expect the same kind of reaction from others as that provoked by all forms of selfishness.

5. *Social groups have norms that are enforced by punishment.* One of the most important developments in evolutionary biology is the concept of major transitions, which turns fractious groups of organisms into cooperative units. A major transition occurs when selection within groups is suppressed, causing selection among groups to be the primary evolutionary force. This concept was first proposed in the 1970s to explain the evolution of nucleated cells as symbiotic communities of bacterial cells.[7] Then it was generalized in the 1990s to include such major transitions as the evolution of multicellular organisms, social insect colonies, and perhaps even the origin of life itself as groups of cooperating molecular interactions.[8]

Even more recently, it is becoming apparent that human evolution qualifies as a major transition. Our ancestors became the primate equivalent of a beehive, achieving ecological dominance through their ability to cooperate. Cooperation includes *mental* in addition to *physical* activities. Most of the mental abilities that set us apart from other species, such as the capacity for symbolic thought and extensive reliance on socially transmitted information, are forms of cooperation that require a degree of trust among social partners.[9]

Earlier I said that a major transition involves the suppression of fitness differences within groups, causing selection among groups to be the dominant evolutionary force. In our species, the suppression of fitness differences within groups is accomplished largely through norms that are enforced by punishment. Our capacity to create norms, largely abide by them, and punish those who don't is so deeply ingrained that most of the time it operates beneath conscious awareness.

Before discussing serious norm violations, I'd like to tell a story about a trivial violation I myself committed. I am not a bowler, but one day I decided to go bowling with my family. I was unaware of a bowling norm, which is to wait if the person in an adjacent lane is about to bowl, before taking one's own turn. After I had violated this norm a few times, an employee kindly and discreetly told me what I was doing wrong. Even though my "punishment" was as gentle as it could possibly be, I was so ashamed that I became physically incapacitated. My bowling score, already low, plummeted even further and I fled the establishment as soon as I possibly could—to the comfort of my accustomed activities guided by norms with which I am familiar! In this fashion, most of us perform an elaborate dance guided by norms that are invisible until we make a wrong step. This is not "Evolution 101" because evolutionists are only beginning to appreciate the power and parameters of cultural evolution, something that social constructivists have always emphasized. Nevertheless, the power of norms in human social interactions is on its way to becoming obvious in retrospect from an evolutionary perspective.

The enforcement of norms tends to escalate as needed, beginning with gentle and friendly admonishment, as with my bowling experience, and proceeding to exclusion, torture, and death in extreme cases. Some religions even codify the escalating procedure, as in this passage from a Hutterite text.

> The bond of love is kept pure and intact by the correction of the Holy Spirit. People who are burdened with vices that spread and corrupt can have no part in it. This harmonious fellowship excludes any who are not part of the unanimous spirit. . . . If a man hardens himself in rebellion, the extreme step of separation is unavoidable. Otherwise the whole community would be dragged into his sin and become party to it. . . . The Apostle Paul therefore says, "Drive out the wicked from among you."
>
> In the case of minor transgressions, this discipline consists of simply brotherly admonition. If anyone has acted wrongly toward another but has not committed a gross sin, a rebuke and warning is enough. But if a brother or a sister obstinately resists brotherly correction and helpful advice, then even these relatively small things have to be brought openly before the Church. If that brother is ready to listen to the Church and allow himself to be set straight, the right way to deal with the situation will be shown. Everything will be cleared up. But if he persists in his stubbornness and refuses to listen to the Church, then there is only one answer in this situation, and that is to cut him off and exclude him. It is better for someone with a heart full of poison to be cut off than for the entire Church to be brought into confusion or blemished.
>
> The whole aim of this order of discipline, however, is not exclusion but a change of heart. It is not applied for a brother's ruin, even when he has fallen into flagrant sin, into besmirching sins of impurity, which make

him deeply guilty before God. For the sake of example and warning, the truth must in this case be declared openly and brought to light before the Church. Even then such a brother should hold on to his hope and his faith. He should not go away and leave everything but should accept and bear what is put upon him by the Church. He should earnestly repent, no matter how many tears it may cost him or how much suffering it may involve. At the right time, when he is repentant, those who are united in the Church pray for him, and all of Heaven rejoices with them. After he has shown genuine repentance, he is received back with great joy in a meeting of the whole Church. They unanimously intercede for him that his sins need never be thought of again but are forgiven and removed forever.[10]

Notice that for this religion, at least, the escalating procedure is remarkably humane. The explicit goal is to accomplish a change in *behavior*, not to punish or destroy the individual. Other religions—and other secular cultural systems—are not so humane in punishing extreme norm violations. Either way, for norms that are essential for long-term societal welfare, *something must be done* to prevent them from being violated.

The word "sacred" accurately describes such norms. When something is sacred, you obey its dictates. When something is profane, you use it for your own ends. There is nothing specifically religious about the concept of the sacred, which is why I could invoke it so easily to describe my reaction to Cramer's transgression. When distorting the truth threatens to destroy the entire fabric of society, then the *truth becomes sacred*, or *should* become sacred, for everyone living in the society.

6. *Cultures must provide meaning in a positive sense and work through the emotions in addition to the intellect.* Steven Brown, who studies art from an evolutionary perspective,[11] once pointed out to me in personal conversation that the study of religion from an evolutionary perspective is too intellectual. Real religions achieve their power through stories, rituals, decorated objects, music, and dance. Ellen Dissanayake, who pioneered the study of art from an evolutionary perspective, has even described religion as "a collection of arts activities."[12] This is presumably what the "enchantment" of religion means in a positive sense.

I wholeheartedly agree with Brown and Dissanayake, even though their critique applies as much to my own past work as to that of my colleagues. Cultural systems are not just a list of norms enforced by punishment. They provide a sense of purpose and meaning in a positive sense and are communicated through the emotions in addition to the intellect. The activities associated with the arts superficially seem costly and irrelevant to biological survival and reproduction, which has caused them to be interpreted as nonadaptive by-products of evolution[13] or sexually selected ornaments.[14] Once we appreciate the importance of cultural evolution in our species, however, the adaptive value of the activities associated with the arts becomes manifest. If most of our behaviors come directly from our cultural systems and only indirectly from our genes, then there must be the cultural equivalent of an anatomy, physiology, and replication machinery. From the very dawn of our species, music, dance, stories, rituals, and decorated objects have performed these functions and continue to do so in contemporary societies, religious or otherwise.

*7. Adaptive fictions can remain powerful even when recognized as fictions.* Powerful narratives can provide a strong sense of meaning, even when they make no pretense of describing factual reality. When I read Tolkien's *Lord of the Rings* trilogy and later watched the movies, I didn't need to believe in the literal existence of hobbits, wizards, elves, and orcs to be moved by the message that *power corrupts*. The *message* is factually correct and is a central theme in evolutionary theories of social behavior. The story provides a *vehicle* that delivers the message with a higher emotional impact than would a dry scientific account. The *vehicle* remains powerful even when its elements depart from factual realism, as long as the *message* rings true.

As for narratives in general, so also for religious narratives. As strange as it might seem, literal belief in supernatural agents is not necessarily a requirement for religious belief and participation. One of my favorite examples involves Myles Horton, a great social activist of the early twentieth century, who was born into a devout Calvinist community in the Appalachian Mountains and whose mother was a stalwart of her church. Horton recalls the following conversation with his mother in his autobiography *The Long Haul*:

> One day I went to my mother and said "I don't know, this predestination doesn't make any sense to me. I don't believe any of this. I guess I shouldn't be in this church." Mom laughed and said "Don't bother about that, that's not important, that's just preacher's talk. The only thing that's important is that you've got to love your neighbor." She didn't say "Love God," she said "Love your neighbor, that's all it's all about." . . . It was a good nondoctrinaire background, and it gave me a sense of what was right and what was wrong.[15]

Horton found ample "enchantment" in his nondoctrinaire beliefs. According to Bill Moyers's introduction to *The Long Haul*, "He's been beaten up, locked up, put upon and railed against by racists, toughs, demagogues, and governors." So much for the claim that there are no atheists in the trenches. The best part of the story, however, is that Horton's *mother* could be so clear about the nature of "preacher's talk" and still be committed to her religion.

I suspect that *many* people who are committed to their religious faith are like Horton's mother; they value their religion for its *message*, just as I value *The Lord of the Rings*, and it doesn't bother them that the *vehicle* might not be factually true. Moreover, I suspect that the people who are most aware of the nature of "preacher's talk" are the preachers themselves and others who devote their lives to thinking about religion. Literal belief is most likely to exist at the edges of religion, not at the center, among people who don't have a lot of time and need to be guided by a set of beliefs that can be quickly translated into a set of prescribed behaviors.

## Putting the Parts Together

Now that I have listed some statements that are so basic from an evolutionary perspective—especially when the importance of culture is taken into account—that they are unlikely to be false, I will try to assemble them to draw some conclusions about the prospects for a cultural system that delivers the best of what religion has to offer while fully respecting factual reality.

8. *Everything at the center of religion extends beyond religion.* The most important thing to say about my list of statements is that religion is a fuzzy set—*everything* that defines it also exists outside religion. Another encouraging point is that norms and symbolic belief systems that provide meaning are enormously flexible in their forms and in the behaviors that they motivate. Thus if we think of ourselves as social engineers with the task of designing a secular belief system that emulates the best of what religion has to offer, we have lots of raw material to work with.

9. *Let's not romanticize religion.* When we talk about the enchantment associated with religion, it's important to realize that it is only *sometimes* achieved within religions. It's not as if all religious believers go around enchanted all the time! A recent memoir titled *Closing Time* by social commentator Joe Queenan, who is of Irish descent, describes his life growing up poor in Philadelphia.[16] He experienced the church as a huge part of his culture and life, attending Catholic schools, serving as an altar boy, contemplating the priesthood, and briefly attending a seminary, but there was

no enchantment anywhere, for Queenan or any of his associates, judging from his memoir. Countless other accounts testify to the dull, bureaucratic, ham-fisted conformity-inducing side of religion, not to speak of the deception, backbiting, social climbing, and exploitation that can take place under its cloak. Don't get me wrong—I enjoy studying religions and think that the enterprising social engineer has much to learn from them. My point is that religious cultural systems and nonreligious cultural systems are largely overlapping when it comes to a construct such as "enchantment," if we can imagine measuring it and displaying it as two bell-shaped curves.

10. *Salvaging the reputation of social engineering.* I use the term "social engineering" knowing that it will send chills down the spine of many readers. Past attempts at social engineering have a sad and often sinister history, regardless of whether they were inspired by evolutionary theory, behaviorism, Marxism, capitalism, or any other set of beliefs. When we ask why social engineering so often goes wrong, we arrive at two simple conclusions. First, the attempts are based on faulty theory and faulty knowledge. If you don't understand the problem, you aren't going to find the solution. Second and even more important, social engineering goes wrong whenever it involves some people imposing their will upon others without their consent. People hate being bossed around, and hate turns to loathing when it involves the most important decisions, such as the right to have children. Thus the first rule of social engineering should be that *it must be by consensus.*

Another point to make about social engineering is that there is no alternative. For me, one of the most profound insights of evolutionary theory is that it so often leads to outcomes that are socially undesirable. Socially desirable outcomes *can* result from evolution, including cultural evolution, but *only* under certain environmental conditions, where "environment" is interpreted broadly to include many factors that are potentially under human control. Unless the human social environment is appropriately "tuned," evolution takes us to places that we don't want to go to. If the environment *can* be appropriately tuned, we can achieve outcomes that seemed impossible in the past. In this fashion, evolutionary theory becomes an essential tool for the social engineer. To get started, we must be unafraid to *think* of ourselves as social engineers.

11. *When it comes to evolution, the past is no guide to the future.* It is common to use the past as a guide to what is possible in the future. When we are told that religion is part of all cultures, that war has always

existed, that countless utopian societies have been tried and have failed, and that religious communal societies demonstrably work better than secular communal societies, the implication appears to be that the future must be the same. That implication is dead wrong, at the most basic possible level, from an evolutionary perspective. Evolution is all about the creation of new forms that have never existed in the past.

Not only is a future that doesn't resemble the past an abstract possibility, but it is also a very practical possibility. Let's contemplate what was possible before and after Darwin's theory of evolution. Before, a thoroughgoing atheism was impossible because atheists had no way to explain design in nature. The fact that all cultures have included religions for thousands and thousands of years is irrelevant because something happened exactly 150 years ago to create a new field of possibilities.

The same goes for our understanding of human social behavior. Entire academic disciplines such as psychology, sociology, and economics date back a few centuries at most. They didn't become sophisticated until the twentieth century, and only now are they becoming integrated, with each other and with evolutionary theory, to form a single consilient scientific meaning system. In the previous section I said that social-engineering attempts fail when they are based on faulty theory and faulty knowledge. Past failures are no guide to the future when we consider the trajectory of our knowledge about human social behavior.

Current developments in economics will make these ideas less abstract. Until very recently, economics has been dominated by rational choice theory, an ultrasimple set of assumptions about human nature. The simplicity of the assumptions enabled economists to create a vast theoretical edifice based on mathematical models, as if empirical research or any other source of information about humans were unnecessary. Unfortunately for all of us, the faulty theory and faulty knowledge of economists led their social-engineering attempts to fail.

The first economists to recognize the failure of their own discipline decided that two things needed to be done. First, they needed to perform experiments on people to learn what actually made them tick. Second, they needed to reach outside their own field—to the psychological sciences, for example—for information about what makes people tick. That was the birth of experimental and behavioral economics, which resulted in the award of the Nobel Prize to Vernon Smith (an economist) and Daniel Kahneman (a psychologist) in 1992.

Today, if we use a popular book such as *Nudge*[17] as an indicator, economic theory is reinventing itself based on an understanding of what it means to be a "human" rather than an "econ." I am not suggesting that experimental and behavioral economists are even close to understanding what it means to be "human." In fact, in other articles I am critical of how they have only narrowly consulted psychology and have yet to base their ideas on a firm evolutionary foundation. Nevertheless, if we examine the leading edge of this intellectual movement, we see an intoxicating blend of economists, psychologists, anthropologists, evolutionists, and members of other academic disciplines working together to form a single conception of human social preferences and abilities serviceable enough to guide social policy. It is entirely possible that this more comprehensive knowledge can succeed where previous attempts have failed, as long as it is used to formulate social policy agreed upon by consensus.

Some religious sentiments worth emulating are *humility* and *faith* about the possibility of being *reborn*, or finding *a path to enlightenment* in which individuals and societies can be very different in the future from what they have been in the past. The economists who based public policy on rational choice theory definitely lacked humility. It might seem that I also lack humility in my optimistic assessment of how current and future knowledge can succeed where previous efforts have failed. On the contrary, I think it is possible to cultivate a blend of sentiments similar to the ones mentioned above, where we are appropriately humble about what we know but also confident and optimistic about what we can achieve. If so, then we will have traveled a long way toward "enchantment."

I will end this section with a personal story that involves the editor of this volume. The last time George Levine visited Binghamton University to give a seminar, he stayed at our house, and we enjoyed long conversations late into the night. I doubt that religious enchantment gets much better than that kind of joyous exploration of ideas. When he left, his parting words to me were "The thing I like about you is your sense of *possibility*."

12. *Do something.* In Stephen Crane's famous short story "The Open Boat," he relates the true story of how a ship that he was aboard sank in a storm off the coast of Florida, forcing him to make a harrowing journey to shore in a tiny dinghy with the captain and two other members of the crew, one of whom died in the surf just before reaching land. Remarkably, Crane describes this as the "best experience of my life" because of the profound feeling of love that he felt toward three people who had been strangers just hours before.

This example illustrates the potent effect of codependence on feelings such as love that make up enchantment. Crane's experience can plausibly be interpreted as an innate psychological adaptation that causes people to cooperate when collective survival depends on it, and that is subjectively experienced as love. Millions of Americans experienced the same feeling in the aftermath of 9/11, which made it seem as if their collective survival was at stake. Like Crane, many commented about how *great* they felt, thanks to the feeling of oneness that the disaster had produced. Of course, the feeling was transient and the profiteers went to work almost immediately to make the most of the situation.

Religions make use of our innate psychology when they elicit feelings of codependence, either actual, as when members of religious communities really do depend on each other, or symbolic, as when religious doctrine describes the believer as fighting a cosmic battle of good against evil. Works of fiction such as *The Lord of the Rings* invoke the same feelings without pretending to be real. A belief system that fully respects factual reality is constrained in how it invokes such feelings, but can still effectively do so. In fact, a claim that codependence is real backed by scientific evidence can be more effective than one based on religion, which requires effort to establish its reality.

Codependence probably explains why activists of all stripes experience a form of enchantment close to religious enchantment. When there is a cause and much to do to advance the cause, enchantment comes easily. Thomas Berry, a visionary ecologist, begins his book titled *The Great Work: The Way into the Future* with the sentence "History is governed by those overarching moments that give shape and meaning to life by relating the human venture to the larger destinies of the universe."[18] It should be obvious that Berry is tapping the same source of enchantment as do religions, without invoking the gods.

Conversely, in my experience, many intellectuals and academics suffer from a lack of enchantment because they are unherdable cats who don't *do* anything that involves working with others toward a meaningful shared goal.

13. *Do something locally.* I don't subscribe to evolutionary psychology in all respects, but I do subscribe to the idea that we are genetically adapted to live in small-scale egalitarian societies and are happy when surrounded by trusted associates who aren't trying to boss us around.[19] It follows that doing something locally, such as putting on a bake sale, will be even more

enchanting than doing something that serves a good cause but doesn't include the rewards of actual face-to-face interactions.

I have had an opportunity to experience this for myself during the last few years, based on two new projects. The first is a new think tank called the Evolution Institute, which is designed to formulate public policy from an evolutionary perspective at a national and global scale.[20] The second is called the Binghamton Neighborhood Project, which promotes understanding and improvement of the quality of life in my own community from an evolutionary perspective.[21] Both have the same objectives, but one is global and the other is local. When I work on EI business, I am typically alone, emailing, writing, and so on. When I work on BNP business, I am meeting with people who are also engaged with the community, talking over a beer, or physically doing something such as inspecting a neighborhood. I find the local activities most enjoyable. I also find that the people who are actively engaged with their communities are the happiest people I know, regardless of whether they are religious. Finally, I observe that when a group of people is trying to make a difference together, they don't *care* about their differences in other respects. They don't care that I'm an evolutionist, for example, as long as I'm trying to help, and especially if I can offer useful tools for getting the job done.

When I was researching religions for my book *Darwin's Cathedral*, I was fascinated to discover a social organization called a cell ministry. A large congregation meets as a single body every Sunday, but also meets in smaller groups in somebody's home with food, good cheer, and shared objectives. The design of a cell ministry can be quite sophisticated, with rules about periodically shuffling members of the groups, an unobtrusive monitoring system to ensure participation, and so on. This social organization works well, not because it is religious, but because it provides the psychologically optimal small-scale social environment in its construction of a large-scale social organization.

In my eclectic readings, I also encountered a multicellular society in a completely different place—aircraft carriers![22] An aircraft carrier contains a society of several thousand people who must coordinate their activities with the precision of a Swiss watch. Evidently, they are organized into groups whose members become tightly bonded. Groups even vie against each other in a tribal fashion through sports competitions, but in a way that leads to coordination and solidarity on a larger scale. Brilliant!

It probably won't happen, but I sometimes fantasize about creating an evolutionary cell ministry. Let's say that five hundred people get really interested in the objectives of the Binghamton Neighborhood

Project—understanding and improving the quality of life in our own city from a scientific and evolutionary perspective. In addition to periodically meeting as a whole group, they also meet as smaller groups in someone's home for food, good cheer, and shared objectives. Goodness knows, there are enough problems to go around. One group could focus on prenatal care, another on risky adolescent behavior, another on the design of public spaces. No matter what their day jobs, all participants could acquire the skills of the scientist, becoming sophisticated about the concepts, reviewing the literature, learning about statistics and experimental design, doing research solid enough to be published in peer-reviewed journals, and eventually succeeding at making a difference in a way that has been rigorously assessed for all to see. Would an evolutionary cell ministry work as well as a religious cell ministry? Might it work *better*?

14. *Making the truth sacred.* Cultural systems endure when they do a good job of telling people *what to do*. They are value systems designed to motivate action. In contrast, scientific knowledge by itself does not motivate action. It merely tells us what's out there. It follows that scientific knowledge must interact with a value system for the whole package to function. The value system consults knowledge of what's out there and uses it to inform action. It is important to emphasize the *incompleteness* of science in addition to praising its virtues.

In section 10, I stated that social-engineering efforts fail for two reasons: lack of knowledge and lack of a consensus decision about how to use the knowledge. A consensus decision is essentially an agreement upon values. Thus when we are careful to avoid both failures, we are providing the "whole package" in which science plays an essential but incomplete role.

As discussed earlier, most cultural systems are not organized in a way that carefully distinguishes between a value system on one hand and a body of factual knowledge to be consulted on the other. Instead, values are often communicated in the form of putative facts. If I object to what you are doing, I am likely to call you "sick," "immature," or some other factually incorrect negative to make you stop. If I think that a woman's place is in the home, I am likely to think that women are mentally inferior to men and that it is abnormal for them to have sexual impulses. Facts are so potent for motivating action that establishing something as a fact can be a matter of the highest import, regardless of whether it *is* a fact. Jim Cramer was doing what comes naturally. He was just unusually candid and insensitive to the welfare of others in his drive to establish his own useful fictions.

The merits of treating fiction as fact can be subjected to a cost-benefit analysis. The benefits, which might accrue to either individuals or groups, are the actions motivated by the putative facts. The costs are the consequences of ignoring the real facts over the long term. The belief that global warming isn't caused by people, for example, leads to short-term benefits and potentially disastrous long-term costs. The title of Al Gore's movie, *An Inconvenient Truth*, says it all.

In any given culture, the costs and benefits weigh in favor of apprehending factual reality in some cases and treating fiction as fact in others. That is why all cultures have a mode of thought that can be recognized as rational and protoscientific, which is expressed only some of the time, as I stated in section 2.

It is plausible that adaptive fictions are more costly today than they were in the past. Our impact on the planet and our reliance upon technology are simply so great that we must create a sophisticated body of factual information in order to manage them. In other words, there is a greater need than ever before to respect the facts of the matter and avoid treating fictions as facts.

Science, by definition, is a cultural system designed to respect the facts of the matter. The truth is sacred in science and transgressions are punished accordingly, even if the word "sacred" is seldom used. At the end of *Darwin's Cathedral*, I state that I am happy to call science a religion, in which factual knowledge is the god that is being worshipped. Science might work better if it regarded itself in this way.

It is not enough for the truth to be sacred *only* within science, however. If the larger society is dominated by adaptive fictions, then it will not consult science to solve its problems. What's needed is a cultural system that distinguishes values from facts more carefully than ever before. In such a society, the truth would be sacred and transgressions would be punished as they already are within scientific society. Since knowledge by itself does not lead to action, such a society would also include an explicit discussion of values and goals leading to action.

I'm not saying that such a society would be *easy* to implement, but it does seem *possible*. It would go against evolved human nature, since we're hardwired to believe things when they are useful rather than true. Cultures *already* go against evolved human nature, however, as when religions establish impossible ideals of altruism and saintliness that can only be approached and hardly ever achieved. Moreover, we already have an example of a culture that successfully treats the truth as sacred—science. We merely need to expand the scientific ethos about the sacredness of truth, add an explicit discussion of values, and we're done.

This design for society bears an intriguing similarity to therapeutic methods that are proven to be successful for individuals. In Acceptance and Commitment Therapy (ACT), individuals are described as having "chattering brains"—our innate psychology, operating largely beneath conscious awareness, that leads to adaptive outcomes most of the time but can also go awry.[23] ACT encourages an acceptance of one's problems, an explicit exploration of values, and working toward one's true goals rather than fighting the phantoms of the chattering brain. Even a few hours of ACT training have been proven to be effective in well-designed randomized trials. Many of the concepts and techniques employed by ACT are borrowed from religious traditions and are part of the enchantment associated with religion. My suggested design for society can be regarded as ACT therapy writ large.

15. *Scientific knowledge should be adorned with the arts.* Finally, returning to point 6, there is no reason why such a society can't be richly adorned with art that celebrates its values and communicates its content beyond the dry intellectual mode. Three examples will show how this is already happening.

Michael Dowd, an evangelical, was born yet again to become an evolutionist, while remaining a Christian, and he now preaches evolution in the evangelical style with his atheist wife, science writer Connie Barlow. Like the itinerant preachers of old, they are permanently on the road in a van adorned with the image of a Darwin fish kissing a Christian fish. Life *is* stranger than fiction! You can learn more by visiting their Web site, reading Michael's book *Thank God for Evolution*,[24] or purchasing one of his DVDs. I'm here to testify that Michael and Connie pass science journalism standards about the facts while communicating them in a way that adds the enchantment associated with religion.

If the evangelical style isn't to your taste, try Baba Brinkman's *Rap Guide to Evolution*, "a hip-hop exploration of modern evolutionary biology," which is also the name of a stage show that played at the Darwin festival in Cambridge, England, in July 2009, and the highly prestigious Edinburgh Fringe Festival in August 2009.[25] Baba, who has also done *The Canterbury Tales* in rap, did some solid homework in preparation for his evolution show, so he also passes science journalism standards, but the rap medium communicates the content more powerfully than could any professorial lecture. My favorite is a rap based on research by Margo Wilson and Martin Daly on violent behavior in men and early reproduction in women in Chicago inner-city neighborhoods—the culture from which the art form of rap emerged.

My third example is an experience that I had when I was invited to participate in an annual midwinter festival celebrating science and the arts in Ithaca, New York, called Light in Winter. I was paired with an eclectic musical group called Water Bear, consisting of a pianist, a violinist, a cellist, and an electric bassist. The lead member of the group, Mer Boel, had read my book *Evolution for Everyone* and composed music inspired by three themes: individual differences, social control, and expanding the circle of cooperation. The performance consisted of my very short explanations of each theme, followed by their performance while I stood on stage. For the individual difference piece, the musicians adopted shy and bold personalities in their playing style. For the social control piece, the cellist started to dominate the group by playing in Jimi Hendrix style until the electric bassist brought him back into line. For the cooperation piece, the audience was invited to join in a rhythmic chant along with the instruments. The performance had a more powerful impact on the audience than any lecture that I have ever given. I even have a quantitative measure—the audience snapped up *all* of my books available for sale after the event, which seldom happens after my public lectures!

A scientific understanding of humanity can acquire the enchantment associated with the arts, in addition to the enchantment associated with religion. There is much to look forward to when we appreciate that when it comes to evolution, including cultural evolution, the future need not resemble the past.

# 9

## ❧

# Darwinian Enchantment

### Robert J. Richards

Die Geisterwelt ist nicht verschlossen;
Dein Sinn ist zu, dein Herz ist tot!
Auf, bade, Schüler, unverdrossen
Die ird'sche Brust im Morgenrot![1]
—Goethe, *Faust*

M ax Weber, in his darkly elusive essay "Science as a Vocation" ("Wissenschaft als Beruf," 1919), attempted to specify both the purpose and the moral dimensions of modern science. The purpose was stark: to establish the facts of a given matter, and to assess the logical and mathematical relationships among them. Science per se could not—rather, should not—attempt to demonstrate the validity of any moral precepts, nor should the scientist qua scientist clothe a naked world with diverting personal values: "wherever the man of science introduces his own value judgment, there a complete understanding of the facts *ceases*."[2] Weber went further. Modern, progressive science, grounded in intellect and reason, excluded from the natural world spiritual forces, and thus stripped away the enchanting colors of a multidimensional universe, leaving but a bare gray framework. Weberian science was constituted by the conviction that

> one *could*, if one wished, learn about some matter, that, therefore, there were no mysterious, unfathomable forces at play, that one could control all things, in principle, through reasoning [*Berechnen*]. That means: the disenchantment [*Entzauberung*] of the world.[3]

In this essay, I wish to investigate the state of Darwin's world. His theory has been cast by both supporters and opponents as replacing a mind-graced nature with a universal mechanism bereft of moral value;

his theory, they suppose, rendered the modern world disenchanted. George Bernard Shaw, in the preface to his play *Back to Methuselah* (1921), expressed his century's growing fear of the terminating machine of Darwinism:

> You cannot understand Moses without imagination nor Spurgeon [a famous preacher of the day] without metaphysics; but you can be a thorough going Neo-Darwinian without imagination, metaphysics, poetry, conscience, or decency. For "Natural Selection" has no moral significance: it deals with that part of evolution which has no purpose, no intelligence, and might more appropriately be called accidental selection, or better still, Un-natural Selection, since nothing is more unnatural than an accident. If it could be proved that the whole universe had been produced by such Selection, only fools and rascals could bear to live.[4]

I don't believe Shaw is correct in his assessment, and I propose to show that in the Darwinian world one can still hear the sweet birds sing. Darwinian nature, at least as originally conceived by its author, was not denuded of spiritual power or moral value as both Weber and Shaw presumed it was; at least, that is what I will argue in the first part of this essay. But what about our contemporary, neo-Darwinian world? Does it orbit a dying sun and is it itself a dead planet where hope and moral possibility exist only as fossils of an earlier time? Contemporary religious objectors, such as those in the Intelligent Design movement, as well as their opponents among the "New Atheists," all seem to agree with Shaw that Darwin disenchanted the world, and that, if Darwinism prevails, the world will become morally bare and desolate. I believe this also to be a mistaken judgment, and in the second part of this essay I'll attempt to suggest why.

## Whewell's Challenge

A cursory examination of the history of Western science would seem to confirm Weber's analysis. During the Scientific Revolution, the Aristotelian universe was dismantled, at least in part: Newton established gravity as the calculable force governing the operation of material bodies—a force, in his abstemious judgment, that was only a mathematical relationship among bodies. True, Newton still required God as the substrate of space and time and as a benign watchman who occasionally had to adjust the mechanisms of planetary motion. In succeeding centuries, the scope of extrahuman intentional forces further diminished as the natural and social sciences advanced in power and range. Kant and Laplace found resources within Newton's own physics by which to eliminate the need for the God hypothesis in astronomy. At the end of the eighteenth century, demons had been exorcised from dominion over the mad by the new doctors of the mind, Philippe Pinel in France and Johann Christian Reil in Germany, who demonstrated that mental illness was amenable to medical intervention. Yet biology, at least at the beginning of the nineteenth century, seemed resistant to the grinding gears of mechanistic science. Kant had declared that there could be no Newton of the grass blade, and the influential British Kantian William Whewell concurred.

In his *History of the Inductive Sciences* (1837), Whewell demanded strict separation of science from theology.[5] Science operated on the basis of empirical evidence and rational inference, which yielded explanatory laws, whereas theology depended on revelation and hope, which succored a faith in "things not seen." Science explored the mechanistic framework of nature and the principles of its operation; theology unveiled the spiritual forces that erected the framework and authored its laws.[6] From Whewell's perspective, one shared by such eminent biologists as Georges Cuvier and Louis Agassiz, biological organisms manifested teleological properties. In Cuvier's terms, they were subject to the "conditions of existence": internally, the parts of organisms were tightly knit together in means-ends relationships; and externally, the parts of organisms fit into their environmental stations with such precision that any significant alterations of parts—or the environment—would cause extinction of the organism and its type. Whewell agreed: clearly, organisms were purposively designed; indeed, our mind had been so molded in our interactions with the living world that we reflexively analyzed creatures employing concepts permeated with purpose. Nonetheless, as a Kantian, Whewell forbade a leap into the transcendent sphere to explain the designed structure of organisms. Science remained rooted in empirical observation and fixed law, and epistemological strictures confined its operations to the natural world.

The fossil evidence, according to Whewell, did indicate the extinction of ancient organisms and their replacement by progressively higher creatures. But this did not allow any inference of the sort made by Lamarck. Cuvier had shown that over long periods of time no fundamental alteration of species had occurred: mummies of humans, cats, and deer from

Egyptian tombs remained recognizably the same as those living in Paris and in the woods around the city; moreover, the "conditions of existence" would have prevented fundamental species change. Both fact and theory thus argued that "*species have a real existence in nature* and a transmutation from one to another does not exist."[7] Since the scientist could not appeal to scripture for the needed miracles to explain the progressive replacement of species, and since lawful physical causes did not avail, rational inquiry into the origin of species was forestalled. From a scientific point of view, the matter remained "shrouded in mystery, and [was] not to be approached without reverence."[8]

Whewell thus set the problematic for the British biological disciplines in the first part of the nineteenth century: (1) organic life conformed to natural law, while also manifesting teleological, purposive structures; (2) fossil evidence indicated a progressive advance of species over time; (3) but no scientific argument employing natural laws could furnish an account of the intelligent design of organisms or their progressive appearance in the fossil record. The origin of species with their adaptive structures was thus scientifically intractable; only theology might advance a resolution of the mystery.

## Darwin's Response to the Challenge

Darwin had read Whewell's *History of the Inductive Sciences* (1837) a few months after he had returned from his *Beagle* voyage, and he left significant marginalia to indicate that he had carefully considered the older writer's articulation of the problem of a scientific approach to organic life.[9] He undertook the challenge that Whewell had implicitly laid down. In spring of 1837, just before taking up Whewell's *History*, the young naturalist had broached the problem of the origin of species. He believed he could construct a scientific theory that would accomplish what Whewell thought impossible, a theory that relied only on empirical fact and natural law but that nonetheless explained the intelligent design of organisms and their progressive advance. Moreover, Darwin came to believe he could demonstrate what Whewell entirely excluded from the probing hand of biological inquiry: namely, that "man himself, with all his intellectual and moral, as well as physical privileges, has been derived from some creature of the ape or baboon tribe."[10] In the *Origin of Species,* Darwin would construct a theory whose principal task was to explain "the most exalted object we are capable of conceiving, namely, the production of the higher animals,"[11] that is, human beings with their moral instincts.

## The Foundations of Darwin's Theory

After he returned from his *Beagle* voyage in late 1836 and began the job of sorting and cataloging the specimens he had brought back, Darwin became convinced that what he had initially reckoned as three varieties of one species of mockingbird produced by environmental interactions were actually three separate species. Primed with knowledge of his grandfather Erasmus Darwin's transmutational views as well as those of Lamarck, he quickly became convinced that his mockingbirds had been radically transformed. In early spring of 1837, he began to construct a theory of species change. Initially he supposed that each species might have a fixed life span, and that just before one species exhausted its last, another would be born from the old. He quickly came to see, however, that the more likely source of species demise was environmental change. Charles Lyell, the geologist from whom Darwin thought half his ideas had come, had expressed a roughly similar view. Lyell believed that when the environment altered significantly, it would disrupt a species's intimate relationship with its surroundings, resulting in extinction. Lyell suggested that the temporary imbalance in nature would be made good by the appearance of another, replacement species; he hinted that the scales had been readjusted by the divine hand—a suggestion that Whewell would disdainfully dismiss as an effort to slip theological assumptions into science.[12] Lyell did not think of the replacement as an improvement on departed life; indeed, he presumed that in the course of time the same type of creatures would reappear on the earth, that the dinosaurs, which Richard Owen had discovered, would again roam the countryside. Darwin, by contrast, assumed that the replacement of species through geological change would be progressive. As he put it in his *E Notebook* shortly after reading Malthus and initially formulating his concept of natural selection: "My theory certainly requires progression."[13] But when he wrote that, his theory was mostly scattered ideas and uncertain aspirations.

Through the last half of 1837 to the fall of 1838, Darwin proposed two forces that might cause the alteration and progressive development of species: the direct effects of the environment and inherited habit. The latter device might explain, for example, how certain waterbirds acquired their webbed feet: the habit of swimming out on the water to feed would, after generations of practice, become instinctual; and such instinct, in its turn, would require the stretching of the birds' toes, which would eventually produce an extension of skin between the toes. He thus concluded: "All structures either direct effect of habit, or hereditary & combined effect of habit."[14] Darwin thought this mode of account, because it did not attribute anatomical change directly to the actions of conscious will, to have escaped the usual objections directed against Lamarck. Of course, we can recognize a good deal of protective defense in Darwin's effort to put distance between himself and his much-maligned French counterpart.

In late September 1838, Darwin began reading "for amusement," as he said, Malthus's *Essay on the Principle of Population*.[15] Malthus planted the seed of what would become Darwin's device of natural selection. This new causal principle would push into the background his other, Lamarckian devices, though without eliminating them from his explanatory repertoire. The usual understanding of the historical development of Darwin's conception runs as follows. Darwin had already been aware that the breeder's "picking" could transform the structure of animals and plants, but initially he had no way of conceiving how such a process might occur in nature.[16] Malthus provided the key notion of population pressure: many more offspring of organisms—whether they be daffodils or elephants—would be produced than the environment could sustain. Hence if among the individuals so produced, any had a trait that gave them a slight advantage, they would have a better chance of reaching reproductive age and passing on that trait. Gradually, then, members of a species would change through the slow accumulation of adaptive advantages. This historical scenario is not wrong, but it overlies the deeper assumptions and motives of the theory, those hidden powers holding the theory together. Below the surface lay the still-enchanted world that Darwin never quite abandoned.[17]

## The Purposive Structure of Nature

In the Malthus moment, when Darwin initially considered the effects of population pressure on organisms, he likened the impact to "a hundred thousand wedges trying [to] force . . . every kind of adapted structure into the gaps . . . in the oeconomy of Nature, or rather forming gaps by thrusting out weaker ones." Immediately, though, he sought to understand the event in teleological terms, terms of the sort that Whewell had thought endemic to biological understanding. He quickly jotted: "The final cause of all this wedging, must be to sort out proper structure & adapt it to change—to do that, for form, which Malthus shows, is the final effect, (by means of volition) of this populousness on the energy of Man."[18] Thus as Darwin initially construed the operations of what would become his device of natural selection, he understood the purpose, or final cause, of population pressure to be alteration of organic form, and he suggested that such purpose might be compared to what humans intended in similar circumstances.

The appeal to a final cause here might be thought merely a *façon de parler*, something the careful historian should not take very seriously. After all, many scholars have contended that Darwin's new theory completely banished from modern biology teleological reasoning about nature.[19] It is nonetheless clear that the telic mode of consideration permeated Darwin's early construction of his theory and, I believe, shaped the expression of that theory in the *Origin of Species*. For example, when he contemplated the virtually limitless periods of time that Lyell projected for the formation of the earth's geological structures, Darwin immediately cast that fact into a purposive account: "Progressive development gives final cause for enormous periods anterior to Man."[20] That is to say, the great expanse of time antecedent to the present could be explained as a requirement for the progressive development of human beings with their distinctive faculties. Or in early November 1838, when tackling a problem that still intrigues biologists—why is there sexual generation instead of the more simple asexual modes—Darwin construed that, too, as requiring a teleological explanation:

> My theory gives great final cause of sexes: for otherwise, there would be as many species, as individuals, . . . if all species, there would not be social animals . . . which as I hope to show is the foundation of all that is most beautiful in the moral sentiments of the animated beings. If man is *one* great object, for which the world was brought into present state.—& if my theory be true then the formation of sexes rigidly necessary.[21]

This passage expresses an explicitly teleological account of the antecedent presence of sexual generation: it was for the purpose of producing social animals; and social animals came to exist for the purpose of ultimately producing moral animals, namely, us. At two later times, Darwin went back over this passage, inserting a qualification on the "great final cause" of the first line. More cautiously, he remarked: "I do not wish to say only cause, but one great final cause. . . ." And still later, perhaps when composing the *Origin,* he further qualified the "*one* great object" of the last sentence: "although, that it was the sole object, I will dispute, when I hear from the geologist the history, & from the Astronomer that the moon probably is uninhabited." Despite these qualifications, the structure of teleological reasoning in Darwin's considerations was preserved right up to the composition of the *Origin of Species.*

There were many other instances in Darwin's notebooks when teleological analyses of particular phenomena were explicitly proffered.[22] And if one added the many instances in which Darwin employed "purpose"— or its more obscure synonym "object," as in his remark above about "*one* great object"—then both the notebooks and the *Origin* are rife with such language. "Purpose" or "object" occurs some sixty-three times in the *Origin,* while "mechanical," "mechanistic," or any of its forms occur only five times—and none modifying "natural selection." The belief that Darwin's theory banished final causes or the application of purposes appears, then, rather exaggerated.

There is one use of "final cause" that Darwin does repudiate: when a purposive trait is ascribed to the direct action of the Deity instead of to the operations of natural law. In his *M Notebook,* Darwin observed:

> This unwillingness to consider Creator as governing by laws is probably that as long as we consider each object an act of separate creation, we admire it more, because we can compare it to the standard of our own minds, which ceases to be the case when we consider the formation of laws invoking laws. & giving rise at last even to the perception of a final cause.[23]

Through the notebooks and his essays of 1842 and 1844, and into the *Big Species Book,* which would be abridged and completed as the *Origin of Species,* Darwin argued that events in nature had to be understood as occurring through natural law. That certainly was the message of Whewell's *History of the Inductive Sciences.* But how, exactly, did natural law manifest itself in Darwin's theory?

## Natural Selection as Natural Law

In the last paragraph of the *Origin of Species,* Darwin specified by way of summary the laws that he discriminated in his book. They included "Growth with Reproduction," "Inheritance," "Variability," "Struggle for Life," and "Natural Selection." In the *Descent of Man* he referred without hesitation to "the law of natural selection."[24] And in his *Autobiography*, he contrasted his law of natural selection with Paley's interventionist deity:

> The old argument of design in nature, as given by Paley, which formerly seemed to me so conclusive, fails, now that the law of natural selection has been discovered. We can no longer argue that, for instance, the beautiful hinge of a bivalve shell must have been made by an intelligent being, like the hinge of a door by man.[25]

In his notebooks and in the *Origin*, Darwin would contend that the creation of new species occurred by law, the law of natural selection. But what exactly did he mean by law, and how did natural selection operate as law?

By law, Darwin seems to have meant causal interactions in the natural world that were fixed and of an unchangeable type. These interactions formed a network of radiating forces that governed all inorganic and organic formations. The most general physical causes, he suggested, had a determining impact on a more specific range of causes, and these in turn were translated into environmental alterations—for example, slow geological changes—that gave shape to organic adaptations. He thus could have almost agreed with the Aristotelian dictum that man was generated by man and the sun, so hierarchical a nexus did he propose. Darwin considered this conception of a network of laws shaping organisms to be quite superior to the "cramped imagination that God created (warring against those very laws he established in all organic nature) the Rhinoceros of Java & Sumatra, that since the time of the Silurian, he has made a long succession of vile Molluscous animals."[26] A universe of fixed physical and organic forces precluded any idea of free human will, and Darwin was certainly not a partisan of that presumption. His conception of a universe of fixed forces determining all events and even human behavior appears to have hastened the rise of the disenchanted modern world. This impression, though, is mitigated both by his assumption concerning the ultimate cause of law itself and by his conception of the operations of the principal law of organic life, natural selection.

As the passage just quoted suggests, Darwin assumed a view of natural law quite common in the early nineteenth century, namely, that law by its very nature required a mind to formulate it and provide the power to enforce it. William Paley, in his *Natural Theology*, expressed this general view:

> A law presupposes an agent, for it is only the mode according to which an agent proceeds; it implies a power, for it is the order according to which

that power acts. Without this agent, without this power, which are both distinct from itself, the "law" does nothing; is nothing.[27]

William Whewell made a comparable assumption, which for him meant that natural law could be assigned the creative process in nature; it could act as a surrogate for God. Whewell put it this way in his *Bridgewater Treatise*, in a passage that Darwin used as an epigraph for the *Origin of Species*:

> But with respect to the material world, we can at least go so far as this—we can perceive that events are brought about not by insulated interpositions of Divine power, exerted in each particular case, but by the establishment of general law.[28]

Like Whewell, Darwin believed that the creative power of nature, and thus the explanatory power, lay in natural law. In the manuscript of the *Origin of Species*, he simply defined nature as "the laws ordained by God to govern the Universe."[29] And as he put it to Asa Gray, a supporter in America: "I am inclined to look at everything as resulting from designed laws, with the details whether good or bad, left to the working out of what we may call chance."[30] By the time he wrote Gray in spring of 1860, however, Darwin had begun to waver in his conviction that natural law required an independent designing mind to provide its force. And by the end of the 1860s, he seems to have abandoned altogether the idea that God was a necessary foundation for his theory. But what he seems never to have abandoned is the ascription to natural selection itself of those properties of discrimination, power, and moral concern previously conferred on it by divine agency. These properties allowed the law of natural selection to lead to the end Darwin foresaw as the goal of the evolutionary process, an outcome that Whewell thought impossible, namely, the natural creation of man as a moral creature.

## The Dual Aspect of Natural Selection

It is easy to assume that the idea of natural selection sprang from Darwin's head full-blown, with all the features we commonly attribute to it. But the evidence of his notebooks suggests a more gradual development of the idea, a development that had to overcome many conceptual obstacles, difficulties that become more apparent when we examine the two major features of the idea of natural selection: struggle leading to survival of the individual, with the consequent preservation of advantageous traits; and the mating of those organisms having favorable variations, and so the transmission of those traits to descendents, thus modifying species over time. Darwin had to put these two parts together, which I believe he never successfully did.

In 1842, Darwin drew up a thirty-five-page pencil sketch of his emerging theory, and two years later he expanded this outline to over two hundred pages. These two essays became the template for the *Origin of Species*.[31] In the essays, he sought to apply the model of domestic breeding to organisms in nature. In the domestic situation, new and variable environments would have an impact on organisms, particularly on their reproductive systems. These environmental impacts would, Darwin believed, ultimately produce smaller or larger variations in the progeny of the affected breeds. Those in the wild state would also be subject to comparable environmental changes, though occurring more gradually and at a slow geological pace. "Hence we should expect every now and then a wild form to vary."[32] He contended that Malthusian population pressure would produce a struggle giving the advantage to organisms with slightly more favorable traits. "In the course of a thousand generations infinitesimally small differences must inevitably tell."[33] The result would be the morphological change of species. Darwin yet recognized a residual problem, which he posed in his early sketch:

> Is there any means of selecting those offspring which vary in the same manner, crossing them and keeping their offspring separate and thus producing selected races: otherwise as the wild animals freely cross, so must such small heterogeneous varieties be constantly counter-balanced and lost, and a uniformity of character [kept up] preserved?[34]

The problem that Darwin recognized may be called the swamping problem: favorable variations will occur only occasionally to individual organisms; in the domestic situation, breeders will select and segregate from large flocks just those few individuals with desired traits and allow them to mate. But what selects and segregates organisms for breeding in the wild and keeps them from crossing back into the unmodified or negatively modified individuals of the group? What power brings such selected organisms together for mating and thus creates a new, transmuted line? In the domestic situation, it's the breeder who selects and segregates a few individuals from a large flock and mates them. But what power performs this function in nature? In both the 1842 and 1844 essays, after Darwin had broached the problem, he introduced a model for what occurred in the wild:

> Let us now suppose a Being with penetration sufficient to perceive differences in the outer and innermost organization quite imperceptible to man, and with forethought extending over future centuries to watch with unerring care and select for any object the offspring of an organism produced under the foregoing circumstances; I can see no conceivable reason why he could not for a new race (or several were he to separate the stock of the original organism and work on several islands) adapted to new ends. As we assume his discrimination, and his forethought, and his steadiness of object, to be incomparably greater than those qualities in man, so we may suppose the beauty and complications of the adaptations of the new races and their difference from the original stock to be greater than in the domestic races produced by man's agency.[35]

Darwin here has formed for himself a model of the selector operative in nature, a selector with preternatural "forethought" and "discrimination," who chooses organisms because of their "beauty and complications of adaptations," and does so with "unerring care." Like the domestic breeder, this imaginary being would segregate individuals with favorable traits and prevent backcross to the rest of the group. This same power inhabits the description of natural selection in the *Origin of Species*.

In his book, Darwin compared the discerning and penetrating actions of natural selection with the careless and superficial actions of the human breeder:

> Man can act only on external and visible characters: nature cares nothing for appearances, except in so far as they may be useful to any being. She can act on every internal organ, on every shade of constitutional difference, on the whole machinery of life. Man selects only for his own good; Nature only for that of the being which she tends. . . . Can we wonder, then, that nature's productions should be far "truer" in character than man's productions; that they should be infinitely better adapted to the most complex conditions of life, and should plainly bear the stamp of far higher workmanship?[36]

In the *Origin*, Darwin retained the model of selection that he had formulated in the essays, that of a wise and morally concerned agent. But what was the analogue in nature for this model? It was, of course, the "struggle for existence."

Yet even the notion of "struggle for existence" was, as Darwin recognized, a metaphor. It encompassed in a literal fashion two dogs struggling over a piece of meat as well as a plant struggling for moisture at the edge of the desert. It applied to plants struggling to produce fruit that would tempt birds, as well as plants struggling with others to germinate the most seeds.[37] The result of these various kinds of struggle would be, as the subtitle of his work put it, "the preservation of the favored races." Yet was it the "preservation" of the fit or the "elimination" of the unfit?—a much-debated question in the later part of the nineteenth century. As we understand natural selection today, no individuals are actively preserved—they simply haven't been eliminated and so live to procreate and pass on their traits. Preservation implies some active intervention by a solicitous hand. But likewise, elimination also suggests an intervening hand, a deadly one that exterminates—or at least prevents the breeding of—some varieties. Even when Darwin attempted to characterize what the process might be in nature that his model indicated, he still described it as if an intentional agent were at work.

In the *Origin,* Darwin noted that successful breeders kept large flocks, which would increase the chance of favorable variations appearing.[38] He argued analogously for what occurred in the state of nature: natural selection would operate more swiftly and powerfully in large open areas where "there [will] be a better chance of favourable variations arising from the large number of individuals of the same species."[39] Under those conditions the discerning eye of natural selection might pick out favored organisms. What Darwin simply assumed was that natural selection, just like the attentive human breeder, would segregate such individuals for mating. Most current biologists recognize that large numbers of a species would only exacerbate the swamping problem. That is why today most ecologically minded scientists believe evolution under selection will occur more rapidly in small geographically isolated spaces—on islands, mountainsides, lands sequestered by natural barriers—where only a small number of individuals might dwell. Darwin asserted the advantages of sympatric speciation because of his *agential* view of natural selection; contemporary biologists think allopatric speciation is required, because of their interpretation of natural selection as a *mechanical* process of change.

## The Moral Character of Natural Selection

The power of Darwin's model of natural selection as discerning mind can be assessed in light of another feature of his analysis. This mind, as

Darwin characterized it in the essays and in the *Origin of Species*, cared for the organisms that it tended:

> It may be said that natural selection is daily and hourly scrutinizing, throughout the world, every variation, even the slightest; rejecting that which is bad, preserving and adding up all that is good; silently and insensibly working whenever and wherever opportunity offers, at the improvement of each organic being.[40]

Darwin here asserted that natural selection worked for "the improvement of each organic being." Immediately following that passage he reiterated: "natural selection can act only through and for the good of each being."[41] In the penultimate paragraph of the *Origin*, he again affirmed the moral concern that natural selection evinced: "And as natural selection works solely by and for the good of each being, all corporeal and mental endowments will tend to progress towards perfection."[42] These are not slips of the pen, since he made the same assertion several other times in the *Origin*.[43] But, of course, from our perspective, natural selection does not work for the good of each being; it eliminates most beings, destroys them. I believe Darwin's conception of a benevolent mind operating in nature had such deep roots in his theory that it overcame what appears to be, at least for us, an obvious consequence of the actions of natural selection. In those brief moments when the patent logic of the situation did hit him, he found ways to assuage the consequences:

> When we reflect on this struggle, we may console ourselves with the full belief, that the war of nature is not incessant, that no fear is felt, that death is generally prompt, and that the vigorous, the healthy, and the happy survive and multiply.[44]

Even here, Darwin suppressed what he had otherwise maintained, that natural selection is "daily and hourly scrutinizing throughout the world every variation"—natural selection did act constantly; the war of nature was incessant.[45]

Not only did nature, in the form of natural selection, exhibit moral concern, it had the goal of producing moral animals, that is, human beings as the "*one* great object for which the world was brought into present state," as Darwin put it in early November of 1838.[46] It was, then, incumbent on him to work out just how a moral sense might have evolved in man, which is what he set out to do in early October of 1838, when he began a new notebook devoted to morals and mind, his *N Notebook*.[47] That human beings were "one great object" explained for Darwin, as we have already seen, why sexual generation came into the world: sexual generation produced social animals, which were the only sort that could have social instincts, and social instincts were, he believed, the foundation of moral behavior. In his *N Notebook* and in associated loose notes, he initially presumed that social habits of parental care, solicitude for offspring, and cooperation among group members would gradually alter brain structures, rendering such habits instinctual. Thus Darwin looked upon moral impulses, ultimately, as acquired during the course of animal development—not implanted in a soul by God. And just as gravity was a force intrinsic to matter, so "it might with equal propriety be said that the living brain perceived, thought, remembered &c."[48] Yet, as he considered the subject, he judged that his kind of materialism did not imply "atheism," a charge made against Lamarck's comparable theory. Darwin cast his materialism in an ennobling teleological framework:

> This Materialism does not tend to Atheism. Inutility of so high a mind without further end just same argument. Without indeed we are step towards some final end.—production of higher animals—perhaps, say attribute of such *higher* animals may be looking back. Therefore consciousness, therefore reward in good life.[49]

Darwin here contended that his view of brain-mind did not lead to atheism because the sort of material that produced mind had the final purpose of generating the higher animals, that is, organisms with consciousness, moral standing, and thus the capacity for leading a good life with its (eternal?) reward. As he put it a few years later, in his essay of 1844, the developmental process he traced led to "the most exalted end which we are capable of conceiving, namely the creation of the higher animals."[50] That, of course, was also the expressed goal he specified in the last paragraph of the *Origin of Species*.

## Darwin's Moral Theory

Darwin's construction of a theory of morality remained, in these early years, fairly rudimentary. Other-regarding behavior would be the result of inherited habit. Initially, he did not see his way clear to applying natural selection to behavior. And when he tried to do so in the 1840s, he met a severe obstacle. While reading in the literature of instinct, he recognized that colonies of ants and bees had workers with distinctive behaviors and anatomical structures; some species of ant, for instance, might exhibit three different castes of workers distinguished by their instincts. Natural selection operated on individuals that had advantageous traits to promote their survival and allow them to pass those traits to offspring. But workers in social insect colonies were neuters; they produced no offspring. Darwin hit upon this problem in the 1840s, and found no easy way to bring such instinctual behavior under the aegis of natural selection. Indeed, he thought this problem might be "fatal" to his whole theory.[51]

Darwin did find a solution even as he worked on the "instinct" chapter of the *Origin of Species*. Natural selection would operate on the whole nest or colony, giving the advantage to those groups that had by chance individuals that acted cooperatively or protectively. This opened up the possibility to consider how moral behavior, which is usually directed to the advantage of another instead of self, might have evolved. And in the last part of the 1860s, he constructed a well-articulated theory of the evolution of the moral sense in man. Like the social insect species, proto-humans, Darwin supposed, began their trajectory toward full humanity in small tribal groups and clans. As he said in the *Descent of Man*, those groups that had

> many members who, from possessing in a high degree the spirit of patriotism, fidelity, obedience, courage, and sympathy, were always ready to give aid to each other and to sacrifice themselves for the common good, would be victorious over most other tribes, and this would be natural selection. At all times throughout the world tribes have supplanted other tribes, and as morality is one element in their success, the standard of morality and the number of well-endowed men will thus everywhere tend to rise and increase.[52]

Darwin proposed two other principles by which to understand how the altruistic sense might have arisen and have been preserved in communities. He supposed that praise and blame would deter miscreants from taking advantage of the goodwill of their more altruistic brethren—and thus cheaters would be thwarted; reciprocal altruism would also serve to punish noncooperators. But these latter two principles, Darwin thought, would constitute "low motives" for moral behavior.[53] Community selection, by contrast, would instill an authentic morality in a group. He thus judged his theory to have avoided the charge of "selfishness," a charge to which the utilitarian view was liable.[54] He also believed that a progressive intelligence and accumulated learning would be instrumental in teaching individuals that their community extended beyond the narrow tribal group. So that as civilization matured, all men would be looked upon as brothers, members of the same community and thus objects of moral concern.[55]

## The Enchanting Darwinian Legacy: Mind in Nature

Darwin's moral theory, while not protected from the suspicious eye of philosophers in our day, nonetheless provides the grounds, I believe, for an ethics that meets the standards for a normative system. Moreover, empirical work in psychology and anthropology suggests that the evolution of altruistic response might well constitute the foundation of modes of ethical behavior displayed by the different nationalities of humankind. That is, Darwin may be understood as having established a universal moral grammar—comparable to Chomsky's notion of a universal linguistic grammar—that grounds all particular moral systems and thus allows us to recognize them precisely as *moral* systems. These, I believe, are defensible propositions, but need to be argued for in another venue.[56] Here I would like to consider what might be the deeper roots for Darwin's conception of the operations of mind in nature.

Darwin acquiesced in the presumption of his contemporaries that natural law implied a lawgiver. His own personal religious beliefs supported that presumption. He had remained a Christian until around 1850; at least that is what he conveyed to Edward Aveling and Ludwig Büchner, who visited him in 1881.[57] And his *Autobiography*, written in the mid-1860s, testifies that when the *Origin of Species* was published, he still believed in "a First Cause having an intelligent mind in some degree analogous to that of man."[58] So during the construction of his theory, Darwin had not abandoned the enchanted sphere; his theory had its foundations well sunk into that ground. Later, in the mid-1860s, when his belief in a supernatural entity slipped to such an extent that he chose to call himself an agnostic, he had not altered the basic structure of his theory to accommodate his release from traditional religion. Alfred Russel Wallace chided him in 1866 for the term "natural selection" since it suggested "an intelligent chooser was necessary."[59] Wallace proposed in its stead Spencer's locution, "survival of the fittest," which did not carry the unwanted implication. But Darwin was reluctant to give up his phrase, though in the fifth edition of the *Origin* (1869), he did title chapter 4 "Natural Selection; or the Survival of the Fittest."[60] So by the mid-1860s, Darwin had relinquished the notion that law required a lawgiver; yet the law of natural selection still retained the capacities that had been transferred to it by reason of its original issuance. Natural selection yet cast the shadow of an intelligent agent. Let me now summarize those capacities that I have already noted and indicate how they differentiate Darwin's principle from our contemporary understanding of natural selection.

First, Darwin's natural selection operates as a *single force*, since it is the outcome of a particular causal law; at least that is the way Darwin expressed its action. Today we would not likely portray natural selection as a law. Moreover, we would regard it not as a singular force, but as the summation of a multitude of different causal forces continuously operating on an organism. Second, Darwin's model of the actions of selection rendered it as an *agent*. The intentional power, as he originally conceived it, came from God; but after he abandoned the assumption of divinity, the causal action of selection still retained those features that allowed the discernment of traits with a refinement that simply could not be equaled by any nineteenth-century machine. In the present day, biologists refer reflexively to "the mechanism of natural selection"—a phrase that never crossed Darwin's lips. Third, natural selection as expressing intentional power not only selected individuals having advantageous traits, it prevented those individuals from crossing back into the general population. Contemporary biologists have different means of handling the problem of backcrosses (aided, of course, by a different understanding of heredity). Fourth, natural selection's intentional character, in Darwin's conception, allowed it the kind of moral solicitude—acting for the good of each creature—that is antithetic to the operations of the modern principle. Fifth, Darwin's device was capable of producing moral creatures, those that might act unselfishly with authentic altruistic impulse. Today both biologists and philosophers tend to be quite suspicious of claims that nature might produce truly altruistic creatures—that judgment, however, is beginning to change. Sixth, natural selection as evincing purpose might produce general progress in evolutionary development. As Darwin put it in the penultimate paragraph of his book: "And as natural selection works by and for the good of each being, all corporeal and mental endowments will tend to progress toward perfection."[61] Most present-day biologists deny that evolution is progressive. Finally, the progress produced by natural selection had as its goal "the most exalted object we are capable of conceiving, namely, the production of the higher animals"—that is, human beings with their moral sentiments. This kind of global teleology is rejected utterly by virtually every evolutionary biologist writing today. The theory issued by Darwin, the theory that forms the foundation of modern biology, had not exorcised the lingering spirit of an enchanted nature.

In this historical account, I have argued that Darwin constructed his theory on the assumption that mind was at work in nature. A salient question remains: does this history have any implications for our contemporary understanding of living nature? I believe that it does. My considerations will perhaps sound a bit too metaphysical for some, but I think they are the result of the soundest common sense, at least if you have been schooled in the works of Kant or, for that matter, William James.

It is quite obvious that human mind formulates the theories that articulate and characterize nature. But more than that. Our daily commerce with the phenomenal world of nature, with its variegated colors, sounds, textures—these all vanish when human mind disappears. The beauty of nature, its delights and terrors—these, too, must be the result of mind active in the world. The moral character of human beings, as well, can only be a function of human mind. These conclusions not only derive from the Kantian perspective, they are the general assumptions of modern biopsychology and anthropology. The world as we immediately experience it is in large part a product of mind.[62] Thus the very ground upon which scientific theorizing rests is already one that is, partly at least, constructed by mind. The scientific depiction of the world, from the Greeks through the accomplishments of the twenty-first century, has changed dramatically. The uncuttable particles of Leucippus and Democritus have been replaced by contemporary theories of the atom, an entity that is entirely cuttable, even down to vanishing mathematical constituents of a sort that Zeno would have delighted in. Despite the changes in the world-image, we assume that the physical world itself, in its basic structures, has remained the same. It is human mind that has changed, and with it the world as we know and interact with it. Our modern world is still inhabited by mind. Shorn of mind, our world ceases to exist.

Little wonder, then, that we still think of organisms and their parts in terms of purposes, of ends. Only mind can hold the past and present together and thus come to understand the activities of organisms in terms of the aims that they display. A flower may achieve its goal in producing the nectar that attracts insects in order that its pollen may be spread to other flowers, but only a mind can come to understand the presence of nectar in the flower in relation to its typical end. What modern biologist would deny that the lens of the eye is designed for the purpose of casting a coherent image on the retina? We now understand such designs as the result ultimately of natural selection. Darwin, though, brought us to discern the capacities of selection because of the original model of mind in nature according to which he conceived the activities of natural selection. Now, particularly as the result of the work of William James, Donald Campbell, and others, we have made the reciprocal analysis: we now conceive mind as if it were a natural selection device.[63] That is, in the effort to solve any problem, we imaginatively fling out variations, possible solutions, until one seems to work—or work sufficiently well for the moment. This kind of blind variation and selective retention grounds most of our inventive efforts at thinking through problems. Darwin may have been moved to construe natural selection in terms of mind because mind, in fact, works like natural selection. For anyone who investigates the process by which Darwin came to construct his theories is well aware that he proceeded by imaginatively trying out this variation and that variation, running down blind alleys until some one of those avenues advanced the solutions he sought. He may well have been dimly aware of his own typical activity in this respect. Thus at a very deep level his own mind may have served as the template for understanding nature. The world remains an enchanted place, though the source of that enchantment may be other than often supposed.

# 10

## ❧

# The Wetfooted Understory: Darwinian Immersions

*Rebecca Stott*

In 1990, only a year after Salman Rushdie had been condemned as a blasphemer by Iranian leader Ayatollah Khomeini and had gone into hiding in fear of his life, he was asked to give the prestigious Herbert Read Memorial Lecture. The novelist's government-appointed security guards could not ensure his safety in such a public arena, so the lecture, entitled "Is Nothing Sacred?" was delivered instead by Rushdie's friend, the playwright Harold Pinter. The lecture, written in isolation, exile, and reflection, and then televised to millions, is an extraordinarily composed indictment of religious fundamentalism and a passionate confirmation of the role of literature in our modern age:

> We have been witnessing an attack upon the very idea of the novel form, an attack of such bewildering ferocity that it has become necessary to restate what is most precious about the art of literature—to answer the attack, not by an attack, but by a declaration of love.[1]

I remember watching Rushdie's lecture with my father, an ex-preacher whose austere fundamentalist religious beliefs had collapsed like a sandcastle in his late thirties, leaving a hole that had filled up with an incoming tide of poetry he'd earlier been taught to shun as worldly (Yeats, Auden, Ridler, Shakespeare, and Rilke). Poetry wasn't a substitute for religious belief for him, he'd insist: it was an entirely new way of seeing. Discovering literature had not been a compensation for the disenchantment of his newly secular world, but rather the first enchantment he had known. He wept during much of Rushdie's lecture. It was as if Rushdie were talking directly to him:

> [Literature] tells us there are no rules; it hands down no commandments. We have to make up our own rules as best we can, make them up as we go along. And it tells us there are no answers; or, rather, it tells us that answers are easier to come by, and less reliable, than questions. If religion is an answer, if political ideology is an answer, then literature is an inquiry; great literature, by asking extraordinary questions, opens new doors in our minds.[2]

Rushdie's lecture addresses many of the questions asked by the writers of this volume but also shapes important new ones. As the lecture unfolds, as if to illustrate his claim that literature, unlike most religions, is dialogic, not monologic, Rushdie brings on from the wings other writers, living and dead, with whom he speaks: Joyce, Melville, Beckett, Cervantes, Calvino, Bulgakov, and Fuentes; the last providing Rushdie with the question that is also at the heart of this essay:

> [Fuentes] poses the question I have been asking myself throughout my life as a writer: *Can the religious mentality survive outside of religious dogma and hierarchy?* Which is to say: Can art be the third principle that mediates between the material and spiritual worlds; might it, by "swallowing" both worlds, offer us something new—something that might even be called a secular definition of transcendence?
>
> I believe it can. I believe it must. And I believe that, at its best, it does.
>
> What I mean by transcendence is that flight of the human spirit outside the confines of its material, physical existence which all of us, secular or religious, experience on at least a few occasions. Birth is a moment of transcendence which we spend our lives trying to understand. The exaltation of the act of love, the experience of joy and very possibly the moment of death are other such moments. The soaring quality of transcendence, the sense of being more than oneself, of being in some way joined to the whole of life, is by its nature short-lived. Not even the visionary or mystical experience ever lasts very long. It is for art to capture that experience, to offer it to, in the case of literature, its readers; to be, for a secular, materialist culture, some sort of replacement for what the love of god offers in the world of faith.[3]

And this is where I pause. Though I agree entirely with Rushdie's claim that literature can, at its best, offer a kind of secular transcendence, I'm not sure about the idea of literature as "replacement." My quarrel is also with the notion of transcendence itself, and that quarrel almost certainly began in my fundamentalist childhood, as a silent questioning of the apparent certainties of my father's sermons. The more austere Christian fundamentalist communities, such as the Calvinist one I grew up in, are notoriously critical of all things of this world. Being fallen creatures, creatures of sin, they preach, all humans are shot through with a kind of moral gangrene. There are no good works that can clear all of that sin up, and no redemption to be had in this world. The only clean, pure place to be is "up there" with the angels—dead and saved. In such communities all eyes are trained perpetually skyward, looking for the signs that the Rapture is about to happen, the promised Rapture in which Christ will return in the night to take his chosen people heavenward. ("Rapture" means ecstatic feeling and an upward transport of mind and body, but as, etymologically, it has also meant fit, abduction, sexual violation, and *rape*, the word holds open powerfully within itself the whole question of consent.) Meantime everything "down here" is blackened and soiled. It's an extreme kind of Manichaeanism of course, but a real one. And there are still plenty of people who believe it and thousands of children who are taught it.

When Rushdie talks about transcendence, then, he doesn't perhaps suspect what I suspect, that the idea of transcendence is kin to that fundamentalist prioritizing of "up there" above "down here," that it might be cousin to that expectation of *uplift*, the longing to be raised up and out of a world we can no longer bear.[4] It is almost impossible to resist metaphors of transcendence; they are, after all, deeply ingrained in our patterns of thought and in our language and metaphors, in the way we talk about the sublime in terms of spiritual revelation: when we hear Mozart's Requiem or a blackbird sing at dawn, or watch the setting sun stain a mountain peak red, we say we are exalted and uplifted; we describe the feeling as rapturous. How do we find a way to describe the often cleaving and searing epiphanies of everyday life when we no longer believe those moments to be god-cloven or god-searing? What words do we use to articulate the new-seeing that comes to us in those moments when epiphanies make us glance not heavenward with wonder and awe but rather earthward with wonder and awe? It is a perpetual struggle for all poets and writers whose business is with the sublime.

For me, as for other writers in this volume, it is not a question of denying that secularism is a form of subtraction from religious "fullness" or of denying that secularism can amount to a "flattening," an emptying out of experience, a disenchantment, but rather a question of finding ways of exposing the religious claim on the powerful experiences that enrich our lives as, in the words of Philip Kitcher in this volume, "the residue of misguided presuppositions that ought to be forsworn." We need to forge a new language for the sublime if we are to forswear those residues and reclaim the sublime for a secular life.

Gillian Beer has shown—brilliantly—that it was a perpetual struggle for Darwin to find a way of expressing his sense of wonder without either reaching for a theistic register or implying a theistic agency.[5] In the nineteenth century, with so few alternatives at his disposal, Darwin struggled to resist casting truth as "up there" and "always beyond." Darwin's struggle is, I've suggested, the *struggle of all secular poets.*[6] In the remainder of this essay, once I have said some things about worms and about Walt Whitman, I want to show how two women poets in the twentieth century—Elizabeth Bishop and Amy Clampitt—found a way of reclaiming the sublime by adopting a poetics of immersion inspired by Darwin. I think they also came to understand something in Darwin's practice as a naturalist that he himself had only glimpsed. Bishop's and Clampitt's search for a new language for the sublime took them back into Darwin's words and methods and into what Clampitt described as the extraordinary "wetfooted understory" Darwin had revealed. They have something to show us there.

## 2

But first, a few words about Darwin and poetry. In a famous passage in his *Autobiography*, written when he was sixty-seven, Darwin declared that he'd lost his taste for poetry.[7] Well, actually, he didn't just declare it, he confessed it. Perhaps he even apologized for it. It's an important passage because it has often been appropriated by anti-Darwinists as a moral billboard to illustrate the supposed flattening-out and deadening of the secular life. *Poor man,* they say. *Look at how he ended up without God. Look at how ill it made him. Look at how sad he became.*[8] But when you look closely, there is actually a great deal more complexity and paradox and humor in the famous passage of the *Autobiography* than has been acknowledged.

> I have said that in one respect my mind has changed during the last twenty or thirty years. Up to the age of thirty, or beyond it, poetry of many kinds, such as the works of Milton, Gray, Byron, Wordsworth, Coleridge, and Shelley, gave me great pleasure, and even as a schoolboy I took intense

delight in Shakespeare, especially in the historical plays. I have also said that formerly pictures gave me considerable, and music very great delight. But now for many years I cannot endure to read a line of poetry: I have tried lately to read Shakespeare, and found it so intolerably dull that it nauseated me. I have also almost lost any taste for pictures or music. . . . This curious and lamentable loss of the higher aesthetic tastes is all the odder, as books on history, biographies and travels (independently of any scientific facts which they may contain), and essays on all sorts of subjects interest me as much as ever they did. My mind seems to have become a kind of machine for grinding general laws out of large collections of facts, but why this should have caused the atrophy of that part of the brain alone, on which the higher tastes depend, I cannot conceive. A man with a mind more highly organised or better constituted than mine, would not I suppose have thus suffered; and if I had to live my life again I would have made a rule to read some poetry and listen to some music at least once every week; for perhaps the parts of my brain now atrophied could thus have been kept active through use. The loss of these tastes is a loss of happiness, and may possibly be injurious to the intellect, and more probably to the moral character, by enfeebling the emotional part of our nature.

Why did Darwin mind so much that poetry no longer excited him? Why did it matter? And why is his lament about lack of feeling so full of passionate feeling?

The passage falls at an important junction in the *Autobiography*. It immediately follows Darwin's description of the last book he had written—on earthworms—a book he had just sent off to the publishers. In describing the worm book, he had caught up with himself in his present moment. He marked the moment by moving from the past tense to the present, and you can almost hear the intake of breath as Darwin writes:

> I have now mentioned all the books which I have published, and these have been the milestones in my life, so that little remains to be said.[9]

But much did *remain to be said*. Darwin knew he was more than the sum of his books, more than those milestones. A further seventeen paragraphs in fact remained to be said as he began to puzzle out for himself what made him unique.

Darwin was only a few years away from death when he wrote this passage in the summer of 1876, and the words he used as he crossed the line from the past to the present and described the decline of his feeling for poetry—*deterioration, dimmed, loss, atrophy, injurious, enfeebling, nauseated*—are the words of a man acutely conscious that his body and mind are failing, and wondering how long he has left to live, or, worse (the worst question of all): *how long will I continue to think coherently?* He felt death to be close (he died only six years later, in 1882). Furthermore, in writing his *Autobiography* in 1876, not only was Darwin *mindful* of his future dead self but—extraordinarily for a man who did not believe in an afterlife—he claimed to be actually writing *as* his imagined dead self: "I have attempted to write the following account of myself," he wrote at the end of that first paragraph of the *Autobiography*, "as if I were a dead man in another world looking back at my own life. Nor have I found this difficult, for life is nearly over with me." [10] Darwin was writing as a dead man, a man who had passed to another world, but, fascinatingly, he was preoccupied in this passage not only with deadness but with feeling.

The strength of Darwin's feeling about the loss of pleasure in poetry in this passage (he describes it, after all, not just as loss of pleasure but in strongly physical terms, as nausea and as intolerance) perhaps also shows what Harold Bloom might characterize as "an anxiety of influence" [11]— Darwin's own necessary (and perhaps guilt-shadowed) attempt to throw off his grandfather Erasmus Darwin's poetic and pseudoscientific legacy. Darwin's feelings about his grandfather were complicated, because though his grandfather's inventions had been praised, his evolutionary ideas and his poetry had in some quarters been vigorously lampooned. Many evolutionary scientists like Erasmus Darwin had been mocked as poets, dreamers, and speculators rather than men of fact since George Cuvier's sustained campaign to discredit Jean-Baptiste Lamarck in the early nineteenth century, [12] but Erasmus Darwin had taken the brunt of the mockery in the English press. Samuel Taylor Coleridge had even coined the phrase "to Darwinise" to describe wild and rash speculation, and he had stigmatized Erasmus Darwin's evolutionary ideas as the "absurd Orang-utang theory." [13] In the nineteenth century, in the British scientific community in particular, speculation had become a dirty word. [14] It was what the French did; it was particularly what Lamarck did; it was what Erasmus Darwin did. As the century progressed, British men of science increasingly defined their practice against such scientific speculation, telling themselves that the British tradition of scientific investigation was based on unbending empiricism, the steady accumulation of facts.

So Darwin had to find a way of exorcising that familial inheritance and minimizing the risk of being lampooned as a mere speculator and a

wild surmiser. He did this, of course, by grounding his theories in fact—
anchoring himself to facts, to this world rather than some imagined one.
He also grounded his work by making networks for himself—connecting
himself up by correspondence with hundreds of naturalists, physicists,
pigeon fanciers, collectors, and natural philosophers around the world
who would vouch for his groundedness. But more strikingly he chose
to ground his theory and indeed his authority as a man of science on
the humble ubiquitous creatures—the overlooked: barnacles, corals, and
earthworms. And earthworms were, of course, of particular importance.
I will return to them later.

One of the most poignant and famous sentences in Darwin's lament
reads: "My mind seems to have become a kind of machine for grinding
general laws out of large collections of facts. . . ." Why *grinding*? What
was Darwin thinking when he chose that word rather than another? It's
a private joke I think—a moment of levity in this otherwise often an-
guished and death-haunted passage. A double joke: one about a schoolmaster and one about worms. The joke is that they are both unflattering
self-portraits. The schoolmaster is Thomas Gradgrind, the hard-hearted,
fact-obsessed, Benthamite schoolmaster of Dickens's novel *Hard Times*,
which Darwin read in 1854 when it was first published, as described in
chapter 2:

> Thomas Gradgrind, sir. A man of realities. A man of fact and calculations.
> A man who proceeds upon the principle that two and two are four, and
> nothing over, and who is not to be talked into allowing for anything over.
> Thomas Gradgrind, sir—peremptorily Thomas—Thomas Gradgrind. With
> a rule and a pair of scales, and the multiplication table always in his pocket,
> sir, ready to weigh and measure any parcel of human nature, and tell you
> exactly what it comes to. . . . he seemed a kind of cannon loaded to the
> muzzle with facts, and prepared to blow them clean out of the regions
> of childhood at one discharge. He seemed a galvanizing apparatus, too,
> charged with a grim mechanical substitute for the tender young imagina-
> tions that were to be stormed away.

Darwin had understood the moral lesson that Dickens had driven home
in that powerful novel: that the fact-grinders are hollow men who must
be redeemed by the circus-people of the imagination. So although Darwin may have suspected as he wrote this passage that it was too late to
redeem his own dulled and still-dulling aesthetic sense, he tells us that if
he could live his life again, he would read poetry and listen to music *at*

*least once a week. Everyone must, he implies, if they are not to atrophy as he has done.*

But there are other grinders in this passage—the worms, the subject of that last great book of Darwin's, *The Formation of Vegetable Mould, Through the Action of Worms, with observations on their habits,* a book being planned and researched and pondered over while he wrote the *Autobiography.* Worms, too, are grinders of matter; they, too, are apparently devoid of feeling, yet they are also highly intelligent and highly adapted. For Darwin the grinding and processing of the worms—incessant and ongoing—provides an alternative and redemptive narrative to the impoverished and impoverishing grinding of Thomas Gradgrind. For, Darwin tells us with admiration, the worms (not God) are the makers and turners and fertilizers of the earth. Grinding is the way they do it.

In those last years of his life Darwin turned to the worms, fascinated by their sense or lack of sense, playing music to them, getting his children to play music to them, and shining lights on them in the middle of the night. He brought them into the house, framed them behind glass in his study, and showed his earthworm experiments proudly to others. In 1880 forty men from the Lewisham Scientific Association and some of their friends were received at Down House. One of them, the Austrian writer Ernst von Hesse-Wartegg, who had written a review of Darwin's book on insectivorous plants in 1875, described the visit in an article in the *Frankfurter Zeitung* a few weeks later. He described two worktables in Darwin's study, one with worms in a glass-fronted wormery, the other bearing the carnivorous plants. He then described Darwin's working methods:

> Darwin has an excellent arrangement, which could be recommended to many another scholar, in the form of a series of wooden boxes, each of which is intended for the reception of all the manuscripts and notes dealing with a particular subject. Thus, for example, the insectivorous plants already referred to, the climbing plants, orchids, domesticated animals etc., together with many other subjects, had their own boxes, which were enriched every day by some note, a newspaper cutting, a relevant object etc. When he is working on some subject 'as now, for example, on earthworms' he then has to supplement and arrange the observations collected over the years in order to complete his work. Of the two dozen wooden boxes, which unconsciously have had such a large part in the production of Darwin's works, a considerable number are already empty. 'They have

done their duty, the material has been processed' and now twenty thousand or even more copies have been distributed throughout the world in book form.[15]

It seems that Darwin's admiring visitor saw the naturalist's work amusingly reflected in the work of the worms and the carnivorous plants who shared his study—just as the worms were grinding and digesting vegetable mold into fertile soil, and the plants were digesting insects, he implies, so was Darwin grinding and digesting facts and ideas into fertile knowledge. Eventually, with the completion of each new book, Darwin's stacked-up boxes of facts emptied out: "they have done their duty, the material has been processed."

Adam Phillips, who writes so eloquently about helplessness in this volume, has argued in his earlier book *Darwin's Worms* that Darwin turned to the worms for consolation for a world of "unredeemable transience," that he sought to show that worms (and not God) had made the earth, and that in the final paragraph of the earthworm book (the last paragraph of the last book—Darwin's last word) he wrote a counterelegiac reprise to the final paragraph of the *Origin* (the famous entangled bank passage).[16]

> When we behold a wide, turf-covered expanse, we should remember that its smoothness, on which so much of its beauty depends, is mainly due to all the inequalities having been slowly levelled by worms. It is a marvellous reflection that the whole of the superficial mould over any such expanse has passed, and will again pass, every few years through the bodies of worms. . . . The plough is one of the most ancient and most valuable of man's inventions; but long before he existed the land was in fact regularly ploughed, and still continues to be thus ploughed by earthworms. It may be doubted whether there are many other animals which have played so important a part in the history of the world, as have these lowly organised creatures. Some other animals, however, still more lowly organized, namely corals, have done far more conspicuous work in having constructed innumerable reefs and islands in the great oceans; but these are almost confined to the tropical zones.[17]

Phillips writes of this passage: "Darwin, in other words, leaves us with a bafflingly simple question the resonance of which he characteristically understates: what would our lives be like if we took earthworms seriously, took the ground under our feet rather than the skies above our heads, as the place to look, as well as, eventually, the place to be?"[18] And this is precisely the point I am reaching for here: the final paragraph of Darwin's earthworm book is not just counterelegiac, as Phillips rightly claims, it is also powerfully *countertranscendent*. It is a commonplace to say that Darwin celebrates the lowly and the overlooked, that it is part of the obvious beauty of a Darwinian way of seeing, but what is less evident is what Phillips points out here—that Darwin would have us look at the earth and the work of the worms, rather than the skies, as the place of miracles. By looking at the skies for meaning and truth, we miss the miracle at our feet. If Darwin had been searching for a way of emptying out the theistic from the sublime, he had found it here with the grinding of the worms.

## 3

Given Darwin's anguish at the disappearance of poetry from his life and his description of himself as having become a machine for the grinding of facts, it is perhaps surprising that so many writers and particularly poets have recognized the poet in him. The naturalist William Tegetmeier praised the poetry of Darwin's work on orchids as early as 1862, quoting Shelley's "Defence of Poetry" as evidence that Darwin was a poet in a Romantic sense:

> In reading this extraordinary work—one which is in every way worthy of the great reputation of its author—the words of Shelley, as descriptive of poetry, continually recurred to mind. "Poetry," says the most eloquent of modern poets, "strips the veil of familiarity from the world, and lays bare the hidden and concealed beauties which are the spirits of its forms;" and in a corresponding passage he describes it as "reducing to union under its light yoke all irreconcilable things." That genius, like poetry, also strips the veil of familiarity from the objects of its pursuit, and lays bare the concealed beauties which are the spirits of their forms, is demonstrated in no ordinary manner by the book under notice.[19]

Several important American poets have recognized the poet in Darwin. In his final years, the half-paralyzed Walt Whitman, whose poetry is full of Darwinian and Lamarckian images and ideas, moved into his brother's house in Camden, New Jersey. There he was visited daily by a young bank clerk and disciple, Horace Traubel, who began making copious notes of their conversations, which he published as *With Walt Whitman in Camden* in nine volumes. Traubel was reading Francis Darwin's newly published *Life and Letters* in 1889 and talked to Whitman about the volumes. On March 21, 1889, three years before Whitman

died, Traubel recorded one of several conversations between him and
Whitman about Darwin:

> When I alluded to Darwin's supposed lack of facility in composition,
> W. said: "I can hardly believe that: I have always thought the opposite:
> thought (from all I had read from him) that he was grandly simple—had the
> sweet directness of a child: that his style was as natural as the bursting of a
> pea in its pod." "And yet that very simplicity has been his power," continued
> W.: "we may say that there is nothing beyond it: it is the enclosing secret."[20]

Two years later, in August 1891, Whitman remarked to Traubel, "I don't
know anything that has gone higher than Darwin—the noble, the exalting. Darwin to me is science incarnate; its spirit is Darwin, Darwin its."[21]
    Whitman's remarks are revealingly inconsistent. On the one hand he
will say that Darwin's writing is sublime, his way of seeing exalting. He
also says that Darwin is science incarnate. He might have used the word
"embodiment," but he chooses instead a word with a heavily theistic
register, "incarnate." He also claims that nothing "has gone higher than
Darwin." He says there is "nothing beyond it," nothing "as high" or better or more profound. It is the "enclosing secret." This is the language
of transcendentalism. Yet only a few years before, commenting on the
"apotheosis" of Darwin after his death, in a short essay on Carlyle in
*Specimen Days and Collect* (1882), Whitman had claimed that evolution
was incapable of reaching the higher truths:

> Unspeakably precious as [the tenets of the evolutionists] are to biology, and
> henceforth indispensable to a right aim and estimate in study, they neither
> comprise or explain everything—and the last word or whisper still remains
> to be breathed, after the utmost of those claims, floating high and forever
> above them all, and above technical metaphysics. While the contributions
> which German Kant and Fichte and Schelling and Hegel have bequeath'd to
> humanity—and which English Darwin has also in his field—are indispens-
> able to the erudition of America's future, I should say that in all of them,
> and the best of them, when compared with the lightning flashes and flights
> of the old prophets and *exalts,* the spiritual poets and poetry of all lands,
> (as in the Hebrew Bible,) there seems to be, nay certainly is, something
> lacking—something cold, a failure to satisfy the deepest emotions of the
> soul—a want of living glow, fondness, warmth, which the old *exalts* and
> poets supply, and which the keenest modern philosophers so far do not.[22]

Whitman's conflict—is the Darwinian vision a high one or a low one? Is
it enchanting or disenchanting? Is it the subject of poetry? Is it the highest

truth ("I don't know anything that has gone higher than Darwin"), or is it only a frustrating reminder of the limitations of our reach, and of how much still remains beyond and unknowable—is typical of the reactions of the metaphysically minded poets who have read Darwin's writing.

Whitman did not resolve his conflict with the high and the low of Darwin.

But Elizabeth Bishop did.

In 1964, Anne Stevenson wrote to the American poet Elizabeth Bishop about surrealism. She was writing a book about Bishop, and she wanted to make a connection between Bishop's poetry and the surrealists' interest in "hallucinatory and dream material" in terms of a shared belief that "there is no split personality [a split between the conscious and unconscious mind], but rather a sensitivity that extends equally into the subconscious and the conscious world." Remarkably, in her letter of reply, Bishop wanted to talk about Darwin, not the surrealists. Bishop wrote to Stevenson:

> Yes, I agree with you. . . . There is no "split". Dreams, works of art (some), glimpses of, unexpected moments of empathy (is it?), catch a peripheral vision of whatever it is one can never really see full-face but that seems enormously important. I can't believe we are wholly irrational—and I do admire Darwin! Reading Darwin, one admires the beautiful and solid case being built up out of his endless heroic observations, almost unconscious or automatic—and then comes a sudden relaxation, a forgetful phrase, and one feels the strangeness of his undertaking, sees the lonely young man, his eyes fixed on facts and minute details, sinking or sliding giddily off into the unknown.[23]

In July of that same year, on a European tour, Bishop came to England intent on seeing "the houses and graves of those who have gone on before."[24] Here she was feted by scores of English poets about whom, in a letter to Robert Lowell, she was scathing: "There is a deadness there—what is it—hopelessness. . . . That kind of defiant English rottenness—too strong a word—but a sort of priggish-ness—as if they've thrown off Victorianism, Georgeianism [sic], Radicalism of the 30s—and now let's all give up together. Even Larkin's poetry is a bit too easily resigned to grimness don't you think? Oh, I am all for grimness and horrors of every sort, but you can't have them, either, by shortcuts—by just saying it."[25] Instead she spent long hours looking at the Piero della Francescas in the National

Gallery, and took a Green Line bus to Kent to visit Darwin's house alone: "a hot summer day (for England) and they were haying all around Darwin's house and it was lovely," she recalled six years later in 1971 in another letter to Lowell.[26] When Edward Lucie-Smith interviewed her for the *Sunday Times* in July that summer, she continued to be "lukewarm" about English poets, preferring instead to talk about Darwin's *Formation of Coral Reefs*. "If you want to read a beautiful book," she told her interviewer, "read that."

Bishop began reading Darwin's *Beagle* journal in January of 1953, exactly a year after she had set up home in Brazil. She was enchanted and quickly resolved to read all of Darwin's work. She continued to return to those books throughout her life. Tantalizingly, she seems to have begun a poem about Darwin around 1962; Lowell saw a fragment of it when he visited her in Brazil that summer. That half-written Darwin poem preoccupied him long afterward: "I am dying to see the ballad and the other new poems. I have been haunted ever since I left Brazil by your fragments and descriptions of the unfinished poems. Hope you've done the Darwin."[27] But though Lowell may have been urging Bishop to finish that poem on Darwin, it seems she never did, for a revolution broke out that summer in Brazil, and by the time she and her partner Lota had escaped to Europe for several months of travel, the moment had passed. Six years later, in 1971 when Bishop was sixty and had been hospitalized after an attack of typhoid fever that left her "limp like a washcloth & unfit for society,"[28] she wrote to tell her friend James Merrill that she had abandoned all books and was reading only Darwin's work: "he is one of the people I like best in the world," she explained.

Remarkably, the duality that Bishop saw and returned to in Darwin—the young man with his eyes fixed on facts "sinking or sliding off into the unknown" in some kind of secular apotheosis or trance—was a duality Darwin himself sometimes struggled to express. While working on the barnacle books in 1850, for instance, he tried to describe to his friend Hooker the constant oscillations he experienced between one way of seeing and another. He wrote, "systematic work wd be easy were it not for this confounded variation, which, however, is pleasant to me as a speculatist though odious to me as a systematist."[29] In other words the philosopher/speculator in him found pleasure in variation because it confirmed his evolutionary ideas (this passage was written while he was working on the barnacles and before his species theory had been published); while the systematizer or fact-grinder in him felt odium for the variation he saw because it made taxonomic classification so frustratingly difficult.

But Darwin had been thinking about and trying to define the nature of his own thought processes for a considerable time before this. In a notebook labeled *This Book full of Metaphysics on Morals and Speculations on Expression—1838* he had tried to analyze the function of dreams and daydreams in the process of discovery. In this notebook he used the term "castle in the air" as a synonym for a daydream or a conscious speculation. It seems that even as early as 1838 he had understood the relationship between the conscious mind and the unconscious mind as a kind of oscillation or slide between reason and imagination.

> I observe a long castle in the air, is as hard work . . . as the closest train of geological thought.—the capability of such trains of thought makes a discoverer, & therefore (independent of improving powers of invention) such castles in the air are highly advantageous, before real train of inventive thoughts are brought into play & then perhaps the sooner castles in the air are banished the better.[30]

Given the antispeculation rhetoric that was prevalent in British science in the nineteenth century, it is not surprising that Darwin was always very circumspect about discussing the role of speculation in public or in print.[31] Such insights tend to occur only in his notebooks or letters. Not surprisingly, what Darwin experienced in the scientific process (the slide between reason and imagination; the slide from the linear train of thought to the "castle in the air" and back again) is visible in his prose. Take this passage, for example, from Darwin's book *The Cross and Self-Fertilisation of Plants* published in the same year as the *Autobiography*; Darwin begins deep in the facts and then begins to speculate:

> We can form no conception why the advantage from a cross is sometimes directed exclusively to the vegetative system, and sometimes to the reproductive system, but commonly to both. It is equally inconceivable why some individuals of the same species should be sterile, whilst others are fully fertile with their own pollen; why a change of climate should either lessen or increase the sterility of self-sterile species; and why the individuals of some species should be even more fertile with pollen from a distinct species than with their own pollen. And so it is with many other facts, which are so obscure that we stand in awe before the mystery of life.[32]

What Elizabeth Bishop recognized in Darwin's writing in that now-famous letter of 1964, and what fascinated her in rereading Darwin during that encounter with her own mortality in 1971, is what I would call the poetics of the commonplace, or what she called in the letter itself "the always-more-successful surrealism of everyday life."[33] If Whitman saw lack in Darwin's work—the lack of the metaphysical vision floating above and beyond—that lack of ascension seemed a refusal and a truth for Bishop. Darwin refused transcendence. Or rather he found the sublime in the material world rather than above it; he did not use the material world as a ladder to something "beyond." And that sense of the wonder and strangeness of what he saw in the plumage patterns of pigeons or the valve mechanism of barnacles or the flight paths of bees created something equivalent to Bishop's "surrealism of everyday life." That giddy slide through the facts into the unknown is, she adds, what the reader should experience in reading poetry, as well as the poet in making it: "What one seems to want in art, in experiencing it, is the same thing that is necessary for its creation."[34]

Had it been possible for them to get to know each other, I imagine Bishop might have put Darwin straight when he expressed his sadness about losing his taste for poetry and lamented that his mind had become a machine for the grinding of facts. The grinding of facts is where poetry begins, she might have said. The sublime is reached through the commonplace, through the slow accretion of facts. Look at what you do, she might have said, look at the way in which you make us slide or sink with you, through the facts, giddily into the unknown.

Here the metaphor Bishop uses is critical—it is a metaphor not of transcendence, but of immersion. You sink or slide through the facts *downward*, or sideways, not upward as part of some transcendent epiphany, some feathered flight. This refusal of transfiguration and transcendence has its roots in Bishop's poetics and in her childhood. Bishop had a conflicted relationship with her forebears, as Guy Rotella points out. She grew up with her maternal grandparents in a Nova Scotia village. Her grandfather was a deacon. Her aunts sang in the Baptist choir, and Bishop described herself as "full of hymns."[35] Her favorite poets were George Herbert and Gerard Manley Hopkins. Yet she was not religious: "she might be called a religious poet without religious faith," Rotella suggests. She also had something of an anxiety of influence in relation to transcendentalism. "I . . . feel that [Robert Lowell] and I in very different ways are both descendents from the Transcendentalists—but you may not agree," she wrote to Anne Stevenson. Rotella argues that although she plays "the correspondence game" like the transcendentalists, reading nature like a book or like a system of signs, searching for significance, the significance she finds is usually secular and earthbound. "She rejects the theocentric, logocentric, and idealist or symbolist epistemologies and esthetics upon which Puritan and transcendentalist doctrines of correspondences are based."[36]

The key is there in the word *sink*. Bishop says that Darwin sinks or slides off into the unknown. If her own poetry is a repudiation of transcendence and indeed of the metaphysical (the notion that all significance is above, *up there*, beyond the physical world), then she recognized Darwin's work as a repudiation of the transcendent in the same way. What she is describing, of course, is the sublime—Darwin's slide or sinking is *giddy*; she describes his sudden awareness of the "strangeness of the undertaking." And it is this erupting, giddying sense of the strangeness of the everyday (a barnacle, a nest, a wasp, a worm) that she finds in Darwin, and that she regarded as "always-more-successful" than the work of the surrealists. "There is no 'split'" between the conscious and the unconscious, between the physical and the metaphysical, she insists, using Darwin as her example—it is a slide or a sinking, not a split. The sublime moment is giddying, but it is not a flight or a transfiguration; it is a fall, a sinking, and an immersion.

Bishop saw something that Darwin was himself aware of—these glimpses of the sublime, the dissolution of self, the moments of wonder and giddiness that, though uplifting and sometimes unavoidably described in a theistic register, were essentially secular and earthbound. The question for both Bishop and Darwin was how to acknowledge the sublime and yet to shake it free of a theistic register. Darwin expressed this himself in a passage from the *Autobiography* that was excised by Emma Darwin in 1887 and then restored by Nora Barlow in her 1958 edition.

> The state of mind which grand scenes formerly excited in me, and which was intimately connected with a belief in God, did not essentially differ from that which is often called the sense of sublimity; and however difficult it may be to explain the genesis of this sense, it can hardly be advanced as an argument for the existence of God, any more than the powerful though vague and similar feelings excited by music.[37]

And that journey through the facts to the point where we slide giddily off into the unknown is there in Bishop's own poetry. It is what she does. "The Moose," for instance, begins by describing a bus meandering along the Nova Scotia coastline "From narrow provinces / of fish and bread and tea" as the sun sets. For stanza after exquisitely meandering stanza Bishop lists the minute details of the landscape and the passengers and the light changing, and the voices whispering through the dusk, until the bus stops with a jolt as a moose steps out of the wood onto the tarmac of the road:

> A moose has come out of
> the impenetrable wood
> and stands there, looms, rather,
> in the middle of the road.

In that abrupt stop and in the exchange of gaze between the animal and the bus passengers that follows it, a collective epiphany takes place in which the passengers, locked in the otherworldly stare of the moose who is "high as a church," slide off giddily into the unknown. It is a Darwinian sublime, a secular enchantment. It begins in the impenetrable wood of the facts and is to be witnessed only by those who attend to the details and to the particular. But it is not an apotheosis. After the searing, giddy slide into the unknown, the bus departs, leaving only the smell of gasoline behind. It is Bishop's glimpse, her surrealism of everyday life. It takes immersion to release the omnipotence of the dream.

In a poem called "In the Waiting Room" written only a year or so after Bishop had claimed she recognized her own practice in Darwin's writing, she describes a child coming-into-consciousness in a dentist's waiting room. It describes the experience she took such pains to define in that letter to Anne Stevenson in 1964, that experience that she saw mirrored in Darwin's writing and which helped her to explain why it was that as a poet she was more like Darwin than like the surrealists. She re-creates the experience for us, taking us through a minutely observed description of the facts of the waiting room and the child turning the pages of the *National Geographic*, until the child slides or sinks into a trance state in which she experiences—stunningly and not without alarm—both her own uniqueness and her kinship with all the other women in the pages of the magazine and in the waiting room:

> Without thinking at all
> I was my foolish aunt,
> I—we—were falling, falling,
> our eyes glued to the cover
> of the National Geographic,
> February, 1918.
>
> I said to myself: three days
> and you'll be seven years old.
> I was saying it to stop

the sensation of falling off
the round, turning world
into cold, blue-black space.
But I felt: you are an I,
you are an Elizabeth,
you are one of them.
Why should you be one, too?
I scarcely dared to look
to see what it was I was.

So Bishop recognized the poet in Darwin in 1964. It startled her, I think. I am not sure she had seen Darwin in her mirror until she started to try to tell Stevenson why she was not a surrealist. She understood, as she knew on some level Darwin did too, the perpetual slide between the known and the unknown, between reason and the imagination, the conscious and the unconscious, in his work, because it was her work too.

## 4

Given Darwin's lament in 1876 that he could no longer respond to poetry, it is surprising that sometime after the publication of the orchid book in 1876, his son William wrote about him: "There was a vague poetic feeling in him. I remember his once saying either about the orchid book or the struggle in nature referred to at the end of the Origin that he almost felt that he could write poetry about it."[38] Francis Darwin also remarkably recalled, "I remember he once said to me with a smile that he believed he could write a poem on Drosera, on which he was then working."[39] What was it about Drosera that made Darwin want to write poetry?

The Drosera or sundew, so named because of the sticky droplets it produces to catch insects, which make it look like dewdrops, is a tiny insectivorous plant. Darwin found one of these plants in 1860 and, fascinated by the extreme sensitivity of its tentacles, brought it home to do experiments on it. By the time he published the orchid book sixteen years later, he had completed hundreds of experiments on those plants, having conscripted them in his search for nervous and digestive similarities between plants and animals.[40] He fed the sundews with milk, urine, saliva, alcohol, and tea, as well as beef, vegetables, and hard-boiled egg, poisoned them with strychnine, quinine, nicotine, and cobra's venom, marveling at how extraordinarily sensitive they were. However, unless there's a piece of paper still waiting to be unearthed by the transcribers and editors of the Darwin-Online Project or the Darwin Correspondence Project, Darwin did not write that poem about the sundew.

But in 1978, an unknown poet called Amy Clampitt did. Born and raised on a farm in Iowa, Clampitt was fifty-eight years old and working as a librarian and freelance editor in New York City when she published an extraordinary poem called "The Sun Underfoot Among the Sundews" in the *New Yorker*. Of course by the time she died sixteen years later, she was very well known, having published six acclaimed collections of poems. These poems are exquisite expressions of what we might call the Darwinian sublime or the poetics of immersion. In this extraordinary poem she asks us to imagine stepping into a bog full of sundews, the bog a metaphor for the lives we lead. She reminds us that we will be swallowed up, that we will not *get out of here*. But there is so much to see, she says, so much light, so much of the sublime. If we look properly, she writes, once we begin to see the sublime beauty here in this Darwinian underworld, in this wetfooted understory, we will begin to *fall upward*.

> An ingenuity too astonishing
> to be quite fortuitous is
> this bog full of sundews, sphagnum-
> lined and shaped like a teacup.
>                              A step
> down and you're into it; a
> wilderness swallows you up:
> ankle-, then knee-, then midriff-
> to-shoulder-deep in wetfooted
> understory, an overhead
> spruce-tamarack horizon hinting
> you'll never get out of here.
>                    But the sun
> among the sundews, down there,
> is so bright, an underfoot
> webwork of carnivorous rubies,
> a star-swarm thick as the gnats
> they're set to catch, delectable
> double-faced cockleburs, each
> hair-tip a sticky mirror
> afire with sunlight, a million

of them and again a million,
each mirror a trap set to
unhand believing,
    that either
a First Cause said once, 'Let there
be sundews,' and there were, or they've
made their way here unaided
other than by that backhand, round-
about refusal to assume responsibility
known as Natural Selection.
                  But the sun
underfoot is so dazzling
down there among the sundews,
there is so much light
in that cup that, looking,
you start to fall upward.

# Notes

## INTRODUCTION

1. In this volume, Philip Kitcher addresses the question of whether Charles Taylor's sense of "fullness" is available to secular people. I agree entirely with Kitcher's analysis that it is, and in the rest of this introduction happily take Taylor's subtle and rich sense of "fullness" as just the condition available to the secular that allows it to fulfill the spiritual needs that religion is usually taken to satisfy. See Charles Taylor, *A Secular Age* (Cambridge, MA: Harvard University Press, Belknap Press, 2007), particularly the opening discussion of fullness, 5ff.

2. Max Weber, "Science as a Vocation," in *From Max Weber: Essays in Sociology*, ed. H. H. Gerth and C. Wright Mills (New York: Oxford University Press, 1946), 155.

3. A very valuable volume discussing secularism in its various manifestations in India and in the West is Rajeev Bhargava, *Secularism and Its Critics* (Oxford: Oxford University Press, 1998). Several of the essays in that volume challenge the idea of secularism that drives this book and the model of the democratic state that lies behind this introduction.

4. Taylor, in *A Secular Age*, offers a history of the way in which we have built "confidence . . . that we as other human beings can sustain a democratic order together, that this is within our human possibilities" (175ff.).

5. In Bhargava, *Secularism and Its Critics*, 31–53.

6. In an interesting and important essay on the relation of secularism to religion, considered in relation to the publication of cartoons of Muhammad that provoked such a violent reaction recently, Saba Mahmood insists that the kind of secular/religious dichotomy with which I am working here is inadequate to a real understanding of the nature of Islamic belief and, of course, any religious belief, and to the nature of secularism itself, which, in its effort to avoid violence in effect imposes on religion contradictory and conflicting norms, and constantly shifts in accordance with the condition of the majority of the population. Part of my point throughout this essay and in giving shape to this book is, however, precisely the complexity of religious faith and the impossibility of reducing it to a set of propositions. Mahmood argues, in agreement with Michael Warner, that the conventional treatment of secularism as "tolerant, satirical, and democratic," as opposed to religious extremism's "uncritical, violent, and tyrannical" character, both caricatures religion and falsely disentangles secularism from its own affective and ideological nature. There is indeed a danger, in making the case for a "democratic" secularism, that religion is being, as she claims, caricatured. And

certainly it is true, as the very rhetoric with which I am here trying to transform secularism implies, that secularism has its own "subjectivity" and "affective attachment." The attempt to claim for this kind of democratic secularism that it is above the fray, capable of rendering something like dispassionate judgment about religious/cultural/political issues, is insufficiently alert to its own limits of perspective and its own history. All of that being the case, and although I am in agreement with Mahmood about the need to establish "a dialogue not only across disciplines but also the putative divide between Western and non-Western traditions of critique and practice," I am comfortable making the case that has generated this book. The effort to alert secularism to its own affective sources, to recognize the power and importance of those sources in religion, and to pursue the questions of epistemology and ontology that divide religion and secularism at the moment seems to me itself to be a secular enterprise. And while it is true that secularism can't be divorced from the ideology of its place and moment, it is sufficient for my purposes that it seek norms that will allow all factions to live in relative harmony together. (Mahmood rightly ends by noting how "the academy" is "one of the few places where such tensions can be explored.") But I do not want to disguise my own commitment, however self-critical it might be, to the value of an epistemology we associate with the growth of Western science. See Saba Mahmood, "Religious Reason and Secular Affect: An Incommensurable Divide?" *Critical Quarterly* 35, no. 4 (Summer 2009): 836–62.

7. *Identity/Difference: Democratic Negotiations of Political Paradox* (Minneapolis: University of Minnesota Press, 1991), 123–57; quotation from 126.

8. Taylor, *A Secular Age*, 5.

9. For my elaboration of this "Darwinian" idea, see my *Darwin Loves You: Natural Selection and the Re-enchantment of the World* (Princeton, NJ: Princeton University Press, 2006).

10. William James, "Is Life Worth Living" (1897), collected in *The Will to Believe* (New York: Dover Press, 1956), 43.

11. Weber, "Science as a Vocation," 142.

12. For a fascinating and untechnical discussion of these questions, see the brief essays by Stanley Fish, "Think Again," *New York Times* online, September 16, 2009, and the book around which the arguments circle, Terry Eagleton, *Reason, Faith, and Revolution: Reflections on the God Debate* (New Haven: Yale University Press, 2009). See also Eagleton's recent *The Meaning of Life* (Oxford: Oxford University Press, 2008).

13. Weber, "Science as a Vocation," 155.

14. James Wood, "God in the Squad," *New Yorker*, August 31, 2009, 79.

15. See Philip Kitcher, *Living with Darwin: Evolution, Design, and the Future of Faith* (New York: Oxford University Press, 2009).

16. See, for a useful collection of Durkheim's related writings, W.S.F. Pickering, ed., *Durkheim on Religion* (Atlanta, GA: Scholars Press, 1994), and, much more recently this time from an evolutionary biologist, David Sloan Wilson, *Darwin's Cathedral: Evolution, Religion, and the Nature of Society* (Chicago: University of Chicago Press, 2002).

17. For a sympathetic but of course critical discussion of this sort of rationalist religiosity, see my "In Defense of Positivism," in *Realism, Ethics, and Secularism:*

*Essays on Victorian Literature and Science* (Cambridge: Cambridge University Press, 2008). In this volume, Kitcher points out the depth and richness of religious traditions and rituals, as opposed to the thinness of the developing traditions of secular humanism, to which his essay (and this book) are pointing.

18. It is fascinating to note how much that tendency to turn "freethought" and science into religion had developed in the nineteenth century. Comte's Positivism, of course, often took the shape of a church, with Sunday sermons; T. H. Huxley and others delivered "Lay Sermons." Karl Pearson's impressive and desperately conciliatory lecture "The Ethic of Freethought" is explicitly determined to affirm that quality of scientific naturalism that insists on its intellectual humility before what is not known (Spencer made it metaphysical with his capitalized "Unknown"), and then affirms the religious nature of the freethinker's mission. The true freethinker, Pearson says, "can relieve a considerable amount of spiritual misery." While he struggles throughout the essay to produce, as Kitcher asks, a secular mission that is profoundly positive, he quickly, almost instinctively, turns the affirmation into a negation, for the "mission" of freethought is "the spread of actually acquired truth—the destruction of dogmatism beneath the irresistible logic of fact." (Pearson, *The Ethic of Freethought* [London: Adam and Charles Black, 1901], 11.) And one need look no further than Bertrand Russell's 1902 essay "A Free Man's Worship" to recognize the quasi-religious mission of the outspoken scientific secularists. (*Mysticism and Logic and Other Essays* [London: Longmans, Green, 1918). These early freethinkers, though often rather "heroic" in their scientific/rationalist stances, were also far more conciliatory to religion and possible demurs than are their current avatars, the outspoken atheists of whom Woods writes.

19. Wilson argues here that the sacred is the realm of ethical requirements, and his title claims, perhaps with a touch of rationalist bravura, that what is sacred is "the truth." Truth and morality are precisely the realms normally associated with religion; bringing them together requires, however, just the secular investment in the processes of this world that Wilson urges in his essay. The more truth, the better the ethical system, or, at least, the more successfully "adaptive."

20. There is no space here to consider this part of the argument, but there are obviously ways in which we know, and have discovered yet more extensively in studies of the psychology of animals, that we are *not* the only minds in the universe. One of the most demanding of questions is what the human relation to other, perhaps less developed minds, ought to be, but it is certainly clear enough that—if without self-consciousness, which *does* seem exclusively human—there are other consciousnesses, some even capable of generosity. The fact of their distinct and almost impenetrable knowing and selfhood creates yet more complex ethical problems.

21. Robert J. Richards, in his essay here and in his earlier work on Darwin, makes a very powerful case, while tracing the development of Darwin's ideas through letters, notebooks, and published writing, that Darwin did not at all begin with this idea of mindlessness, which is Dennett's and modern science's rather ahistorical extrapolation from Darwin's arguments. See, in particular, Richards's *Darwin and the Emergence of Evolutionary Theories of Mind and Behavior* (Chicago: University of Chicago Press, 1987).

22. Adrian Desmond and James Moore, *Darwin: The Life of a Tormented Evolutionist* (New York: Warner Books, 1991).

23. I wish there were more space to talk about the ethical implications of the "lovely." It is important, beyond measure, that the secular be recognized as potentially lovely. Loveliness, of course, implies as well lovability; and there is no possibility of significant secular work in cleaning up some of the messes we humans have made and are making (or confronting some of the ugly things that Darwin and biology reveal to us) unless we can feel the lovability of the world. Connolly makes this case; Jane Bennett movingly claims (one of the implicit themes of this book), "you have to love life before you can care about anything" (*The Enchantment of Modern Life: Attachments, Crossings, and Ethics* [Princeton, NJ: Princeton University Press, 2001], 4. In his essay here, Taylor implicitly approaches the problem of the relation of the aesthetic to the ethical, and gently complains (and it is generously not articulated as a complaint) about the argument of my *Darwin Loves You* that the "enchantment" toward which I point in Darwin's example is primarily aesthetic, like the "wow" with which I have begun this introduction. It is a point well taken. Taylor very usefully associates this sort of aesthetic satisfaction with the world to Kant's notion that such responses to the world are "akin to that we sense before 'the moral law within.'" The aesthetic and ethical interweave around the whole idea of "value," and thus arises the question, so worrying to Victorians and central to philosophy since Hume, of whether one can ever derive an "ought" from an "is," the urgency of value from mere accurate description—ethics from science. It is one of the assumptions and arguments of this book, as we see it suggested by Bennett, that the fullness of aesthetic richness is almost a precondition for the fullness of ethical engagement.

# 1
## Challenges for Secularism

Many thanks to Taylor Carman, Roger Cooke, Wayne Proudfoot, Bruce Robbins, and especially George Levine for comments, discussion, and suggestions.

1. A. C. Grayling has questioned my use of the word "secularism," on the grounds that secularism is a view about the relationship between religion and the state. My choice is based on the need for some term that will cover the views of those who do not believe in transcendental entities. "Atheism" will not do, since it restricts the class of supernatural entities (not all supposed supernatural beings are gods) and also requires denial rather than simple absence of belief (some secularists are agnostics). "Secular humanism" will not do, since many prominent contemporary atheists are, as this essay suggests, light on the humanism. Moreover, my use of the term is hardly idiosyncratic: to cite just one example, in her magisterial biography of Charles Darwin, Janet Browne writes of "Josiah Wedgwood's gradual drift away from the orthodox church towards secularism" (*Charles Darwin: Voyaging* [Princeton, NJ: Princeton University Press, 1996]).

2. William James, *The Varieties of Religious Experience* (in William James, *Writings 1902–1910* [New York: Library of America, 1987]), Lecture II.

3. I use this term not to impute any particular stance on religion to Darwin (his attitudes to religion are matters of scholarly debate), but to empha-

size that many of those who currently campaign for atheism owe and express a debt to Darwin. There are, of course, staunch Darwinians who profess religious faith.

4. See Dawkins, *The God Delusion* (New York: Mariner Books, 2008); Harris, *Letter to a Christian Nation* (New York: Vintage, 2008); Hitchens, *God Is Not Great* (New York: Hachette Book Group, 2007). These books, especially Dawkins's, make some valuable points about forms of religion prevalent in the United States, but, to my mind, suffer from a narrow, and historically uninformed, conception of religion. Dan Dennett's *Breaking the Spell* (New York: Penguin, 2006) is the most sophisticated work in this genre, although it is (in my judgment) insensitive to some important issues. I detail my agreements and disagreements with Dawkins and Dennett in "Militant Modern Atheism" (forthcoming in the *Journal of Applied Philosophy*).

5. Charles Taylor, *A Secular Age* (Cambridge MA: Harvard University Press, Belknap Press, 2007).

6. What I designate here as "the challenge of secularism" is what I previously called "the Enlightenment case against Supernaturalism." See the final chapter of *Living with Darwin: Evolution, Design, and the Future of Faith* (New York: Oxford University Press, 2007).

7. The symmetry argument outlined here is sometimes deployed by sociologists of knowledge to cast doubt on the credibility of scientific claims. See, for example, David Bloor, *Knowledge and Social Imagery*, 2nd ed. (Chicago: University of Chicago Press, 1991). I have tried to argue, at length, that the symmetry can be broken in this instance. See *The Advancement of Science* (New York: Oxford University Press, 1993).

8. Theologians and historians of religion often suppose that the inconsistencies indicate that the scriptural stories were never supposed to be read literally. Hence they are unmoved when militant atheists harp on the incompatibility of various Gospel accounts of the life of Jesus, contending that critiques of this sort apply only to a form of fundamentalism that is historically recent. See, for example, Karen Armstrong, *The Case for God* (New York: Knopf, 2009).

9. Robert W. Funk and the Jesus Seminar, *The Acts of Jesus* (San Francisco: Harper, 1998), 153.

10. The separation can be made in either of two distinct ways. One can take faith to be a legitimate mode of grounding belief, even though the doctrines accepted do not count as items of knowledge. Or one can think of faith as a form of commitment, not expressed in beliefs at all. In this section, I interpret the appeal to faith in the former way. The second approach is implicit in positions I consider in later sections.

11. The discussion of the possibilities of defending against secularism by appealing to blind faith is developed at greater length in my essay "A Pragmatist's Progress: The Varieties of James' Strategies for Defending Religion," in *William James and a Science of Religions*, ed. Wayne Proudfoot (New York: Columbia University Press, 2004), 98–138.

12. Darwinian atheists exult in the ferocity of religious wars, the willingness to torture in the name of the deity (or deities), the intolerant persecution of forms of behavior that are now accepted, and so forth. They pay much less attention

to the positive ways in which religion has entered many areas of individual and collective life. Nor do they take seriously enough the enormities inflicted upon the world by various aggressively secular regimes. A sober evaluation would recognize that the moral track record of secularism, like the moral track record of religious life, is *thoroughly mixed*—that good and bad people have been formed and have acted under both denominations. Perhaps a fine-grained counting could disclose some serious difference, but there are obvious questions about just how the sampling of cases is to be done.

13. John Dewey, *A Common Faith* (New Haven: Yale University Press, 1934), 59. My debt to Dewey in general, and to this short book in particular, is very large, and, in the rest of this essay, I shall be attempting to renew some important Deweyan themes.

14. By this I mean a view of the world that dispenses with supernatural entities. As I shall argue in section IV, accepting a view that is disenchanted in this relatively weak sense does not entail the stronger claim that there is no place for talk of values.

15. See the informed account by Phil Zuckerman, *Society without God* (New York: New York University Press, 2008).

16. Pippa Norris and Ronald Inglehart, *Sacred and Secular* (Cambridge: Cambridge University Press, 2004).

17. This is not, of course, to deny that some churches (and their equivalents for other religions) often block movements that aim at social justice. My point is simply that, for a significant number of oppressed people, religious institutions have offered a route to greater justice and freedom.

18. John Dewey, *The Public and Its Problems* (1927) (Athens: Ohio University Press, 1980). Even earlier, in *Democracy in America*, Tocqueville had emphasized the importance of community structures in North America, and had suggested that there might be difficulties in maintaining them.

19. As I shall eventually maintain, efforts at imitation may initially be pallid. Scientists understand the importance of sustained efforts if experiments are to be made to work. By the same token, we should not be surprised if early attempts to create secular community seem less rich than those that religious groups have developed over centuries, or even millennia.

20. Charles Taylor, *A Secular Age* (Cambridge MA: Harvard University Press, Belknap Press, 2007).

21. Plato *Euthyphro*. In the dialogue, of course, the central terms are different: Euthyphro declares that the pious is what is pleasing to the gods. The modification does not affect the points I make below.

22. Kant recapitulated the point in the *Groundwork*, noting that the idea of arbitrarily willing that certain entities are good makes a mockery of the thought that "the Holy One of Israel" is good.

23. For the pitfalls of programs of this sort, see the final chapter of my book *Vaulting Ambition: Sociobiology and the Quest for Human Nature* (Cambridge MA: MIT Press, 1985).

24. My version of the naturalistic program is presented, fragmentarily and schematically, in a sequence of essays published over the past two decades.

The shape of the account is most easily visible in "Prospects for a Naturalistic Ethics" and "Three Challenges for Naturalistic Ethics." A full version will appear in a forthcoming book, *The Ethical Project*, to be published by Harvard University Press.

25. For Dewey's approach, see *Human Nature and Conduct* (New York: Holt, 1922; repr., New York: Prometheus Books, 2002) and Dewey and James Tufts, *Ethics*, 2nd ed. (New York: Holt, 1932). The last two pages of section II of the latter book (a section authored by Dewey alone) are particularly important in outlining his stance. My forthcoming book, *The Ethical Project*, attempts a full development of that stance.

26. Besides my own efforts in this direction, related attempts have been made by Robert Richards, *Darwin and the Emergence of Evolutionary Theories of Mind and Behavior* (Chicago: University of Chicago Press, 1987); Marc Hauser, *Moral Minds* (New York: Harper Collins, 2006); Frans de Waal, *Primates and Philosophers* (Princeton, NJ: Princeton University Press, 2007); and David Sloan Wilson, *Darwin's Cathedral: Evolution, Religion, and the Nature of Society* (Chicago: University of Chicago Press, 2002).

27. Aristotle's *Nicomachean Ethics*, perhaps the most systematic ancient treatment of the topic, is devoid of any hankerings for permanence. In Plato, however, there are hints of the desire that one's doings achieve a kind of immortality: see, for example, the discussion of two forms of reproduction (those of parents and those of philosophers/educators) in the *Symposium* (207a–d).

28. I believe that the picture of human significance that I sketch very briefly here has been most thoroughly developed in some works of art, music, and literature. In *Joyce's Kaleidoscope* (Oxford: Oxford University Press, 2007), I try to show how it emerges from the explorations of *Finnegans Wake*.

29. Bacon writes, "Men fear death as children fear to go in the dark" (*Essays*, World's Classics [Oxford: Oxford University Press, 1962], 9; I have modernized the spelling). For a sensitive recent discussion of whether death is to be feared, see Julian Barnes, *Nothing to Be Frightened Of* (New York: Knopf, 2008). From the first lines, Barnes self-consciously contrasts his own discussion with the supposedly more rigorous thought of his brother, a professional philosopher, and yet his book is far more attentive to philosophical nuances than many more academic treatments—including those that issue from the word processors of Darwinian atheists.

30. Leonard Huxley, ed., *Life and Letters of Thomas Henry Huxley* (London: Macmillan, 1900), 1:233.

31. The witty intermezzo "Don Juan in Hell," from Shaw's *Man and Superman*, scrutinizes the Christian vision with particular clarity.

32. *A Secular Age*, 5.

33. *The Varieties of Religious Experience*, 379, 380.

34. Bede Griffiths, *The Golden String* (London: Fount, 1994) 9; quoted in Taylor, *A Secular Age*, 5.

35. *A Secular Age*, 5.

36. James Joyce, *A Portrait of the Artist as a Young Man* (London: Penguin, 1992), 157–58. The passage comes almost at the end of chapter 3.

37. Ibid., 159. This is the opening of chapter 4, a less celebrated switch of tone than that in the transition from chapter 4 to chapter 5, but an equally effective one.

38. Dewey, *A Common Faith*, 17.

## 2
### DISENCHANTMENT—REENCHANTMENT

1. "There was a force in all things, animate, and, to our view, inanimate: water, trees, substances, words; and there was a mutual influence among things. There were also human and supra- or extra-human beings, who exercised power of different kinds and at different levels: saints, witches, ghosts, spirits and less palpable entities." Stephen Wilson, *The Magical Universe* (London: Hambledon & London, 2000), xvii.

2. George Levine, *Darwin Loves You: Natural Selection and the Reenchantment of the World* (Princeton, NJ: Princeton University Press, 2006); see especially chap. 1.

3. The inverted commas express some questioning of whether this picture of disenchantment is really Weber's, but we don't need to settle that here.

4. Levine, *Darwin Loves You*, 29.

5. "Zwei Dinge erfüllen das Gemüth mit immer neuer und zunehmender Bewunderung und Ehrfurcht, je öfter und anhaltender sich das Nachdenken damit beschäftigt: *der bestirnte Himme über mir und das moralische Gesetz in mir.*" *Kritik der praktischen Vernunft*, Berlin Academy Edition (Berlin: Walter Gruyter, 1968), 161.

6. Douglas Hofstadter, "Reductionism and Religion," *Behavioral and Brain Sciences* 3 (1980): 434.

7. The awe, but also the sense of connection, emerges in these reflections of Charles Lindbergh: "I know myself as mortal, but this raises the question: 'What is I?' Am I an individual, or am I an evolving life stream composed of countless selves? . . . As one identity, I was born in AD 1902. But as AD twentieth-century man, I am billions of years old. The life I consider as myself has existed through past eons with unbroken continuity. Individuals are custodians of the life stream—temporal manifestations of far greater being, forming from and returning to their essence like so many dreams. . . . I recall standing on the edge of a deep valley in the Hawaiian island of Maui, thinking that a life stream is like a mountain river—springing from hidden sources, born out of the earth, touched by stars, merging, blending, evolving in the shape momentarily seen." He sums up: "I am form and I am formless. I am life and I am matter, mortal and immortal. I am one and I am many—myself and humanity in flux. . . . After my death, the molecules of my being will return to the earth and sky. They came from the stars. I am of the stars." Quoted in Gore Vidal, "The Eagle is Grounded," *Times Literary Supplement*, no. 4987 (October 30, 1998): 6.

We can see how Lindbergh stands fully within the modern cosmic imaginary. His experience of nature, e.g., on Maui, immediately suggests to him the depth of the universe and our dark genesis from it.

8. I have developed this point at greater length in "Explanation and Practical Reason," in *Philosophical Arguments* (Cambridge, MA: Harvard University Press, 1995), 34–60.

9. Bernard Williams, *Ethics and the Limits of Philosophy* (Cambridge, MA: Harvard University Press, 1985).

10. Daniel Dennett, *Consciousness Explained* (Boston: Little Brown, 1991).

11. Merlin Donald, *Origins of the Modern Mind* (Cambridge, MA: Harvard University Press, 1991); *A Mind So Rare* (New York: Norton, 2001).

3
ENCHANTMENT? NO, THANK YOU!

1. Max Weber, *From Max Weber: Essays in Sociology*, ed. H. H. Gerth and C. Wright Mills (New York: Oxford University Press, 1946), 155.

2. Max Weber, *Wissenschaft als Beruf, Politik als Beruf*, herausgegeben von Wolfgang J. Mommsen und Wolfgang Schluchter in zusammenarbeit mit Birgitt Morgenbrod (Tübingen: J.C.B. Mohr [Paul Siebeck], 1992), 109. This is volume 17 of Weber's collected works.

3. Hans G. Kippenberg, *Discovering Religious History in the Modern Age*, trans. Barbara Harshav (Princeton, NJ: Princeton University Press, 2002), 98–99. There is an interesting discussion of this episode in David Morgan, "Enchantment, Disenchantment, Re-Enchantment," in *Re-Enchantment*, ed. James Elkins and David Morgan (New York: Routledge, 2009), 3–22. Morgan's position is "that enchantment, no less than disenchantment, appears inherent to human life" (12).

4. Jane Bennett, *The Enchantment of Modern Life: Attachments, Crossings, and Ethics* (Princeton, NJ: Princeton University Press, 2001), 6. Bennett describes the disenchantment story as follows: "There was once a time when Nature was purposive, God was active in human affairs, human and other creatures were defined by a preexisting web of relations, social life was characterized by face-to-face relations, and political order took the form of organic community. Then, this premodern world gave way to forces of scientific and instrumental rationality, secularism, individualism, and the bureaucratic state—all of which, combined, disenchant the world" (7).

5. Michael Ostling argues that most readers, like J. K. Rowling, take the magic as harmless fantasy. They, like the magic, are "disenchanted" (16). Michael Ostling, "Harry Potter and the Disenchantment of the World," *Journal of Contemporary Religion* 18, no. 1 (2003): 3–23.

6. Paraphrasing Hans Blumenberg, and thus leaving her own views somewhat uncertain, Bennett writes: "The antisecularists are right in that some important source of meaning has indeed been lost with modernity. Call it an ancient cosmos of cyclical events and natural hierarchies, or a world alive with wonders and hints from God, or a teleological substance at work in history and nature—we do indeed live amidst 'the disappearance of inherent purposes.' But this disenchantment is a fact toward which we have adjusted quite well!" (*Enchantment of Modern Life*, 66).

7. Ibid., 63, 7.

8. Fritz K. Ringer, *The Decline of the German Mandarins: The German Academic Community, 1890–1933* (1969) (Hanover: University Press of New England, 1990), 134, 158–59.

9. For example: "Le monde d'hier, tout enchanté qu'il fût, n'était pas constitutivement irrationel, il connaissait ses propres logiques de rationalisation" (Catherine Colliot-Thélène, "Rationalisation et désenchantement du monde: Problèmes de l'interpretation de la sociologie des religions de Max Weber," *Archives de Sciences Sociales des Religions* 89 [1995]: 68).

10. Ibid., 73. The translations are mine. See also Catherine Colliot-Thélène, *Le désenchantment de l'état: de Hegel à Max Weber* (Paris: Minuit, 1992).

11. Terry Eagleton, *The Meaning of Life* (Oxford: Oxford University Press, 2007), 64.

12. Weber, "Science as a Vocation," in *From Max Weber: Essays in Sociology*, ed. H. H. Gerth and C. Wright Mills (New York: Oxford University Press, 1946), 153; hereafter cited parenthetically in text.

13. At any rate, Weber is clearly and forcefully speaking against the "modern intellectualist form of romantic irrationalism" (143)—one that has much in common with the return of religion, explicit and implicit, among intellectuals today.

14. Charles Taylor, *A Secular Age* (Cambridge, MA: Harvard University Press, Belknap Press, 2007), 25; hereafter cited parenthetically in text.

15. As many have noted, the concept of secularization is ambiguous. Does it signal (1) an absolute rupture with religion, or (2) a transfer of religious structures of thought and feeling to nonreligious objects—in other words, a means of keeping religion alive to some degree in some other form? History suggests, for better or worse, that the second option is more frequent. Since the Romantic reaction that followed the Industrial and French Revolutions, the critique of secular rationality has always had a large place in European intellectual life. Thus Vincent Pecora suggests that European and American intellectuals have always had a "love-hate relationship with religion." Much of the critique of European Enlightenment happened, he argues, *within* Europe—and, I think he would want to add, within the Enlightenment. To claim that this critique is internal to the Enlightenment is of course to defend the Enlightenment indirectly by attributing to it the power of self-criticism. Vincent P. Pecora, *Secularization and Cultural Criticism: Religion, Nation, and Modernity* (Chicago: University of Chicago Press, 2006), 111.

16. Hans Blumenberg, *The Legitimacy of the Modern Age*, trans. Robert M. Wallace (Cambridge, MA: MIT Press, 1983).

17. Talal Asad, *Formations of the Secular: Christianity, Islam, Modernity* (Stanford: Stanford University Press, 2003), 2–8.

18. For William Connolly, too, the disenchanted world seems incompatible not with the spirit but with the intensities we experience when we refuse to separate the spirit from the body; "modern secularism" ignores or disparages "the visceral register of subjectivity and intersubjectivity." William E. Connolly, *Why I Am Not a Secularist* (Minneapolis: University of Minnesota Press, 1999), 3.

19. Despite the general disavowal of cultural evolutionism, Vincent Pecora observes, the conflation of secularism with modernity continues to sustain the

terribly misleading idea that everyone must follow the same developmental path, a path on which some are ahead, others behind. "The historical narrative of the advent of secular rationality in the West is almost never told as if it were reversible, as if the West would one day return, for example, to a medieval religious cosmology" (*Secularization and Cultural Criticism*, 29). Making our stories reversible would be, so to speak, a huge step forward. One might say that it is only if secularization were assumed to be reversible that we could properly believe in it as a form of progress.

20. George Levine, *Darwin Loves You: Natural Selection and the Re-enchantment of the World* (Princeton, NJ: Princeton University Press, 2006), 24; hereafter cited parenthetically in text.

21. Slavoj Žižek, *On Belief* (London: Routledge, 2001), 110. It was only after writing most of this essay, and choosing its title, that I read Žižek's book and discovered that its first chapter is titled "Gnosticism? No, Thanks!"

4

SHOCK THERAPY, DRAMATIZATION, AND PRACTICAL WISDOM

1. Hesiod, *Theogony*, trans. with intro. by Norman O. Brown (Englewood Cliffs, NJ: Prentice Hall, 1953), 56, 58, 66, 67, 78, 79. The introduction by Brown is superb. Some will resist this myth on the grounds that it glorifies human violence. But my sense is that while the *Odyssey* and the *Iliad* may do that, Hesiod himself does not. His commendations to Greek farmers in *Works and Days* point toward coping peacefully and wisely with an unruly universe. Others may say that the Olympians smooth out the world of the Titans. Perhaps. But does not Dionysus soon creep in from the East, and was not Sophocles alert to a continuing conflict between these two sets of gods? That being said, I still have much to learn about Hesiod. My goal is not a return but a use. For the essay cited above, see Apostolos N. Athanassakis, *Hesiod: Theogony, Works and Days, Shield* (Baltimore: Johns Hopkins University Press, 1982).

2. Jean-Pierre Vernant, *Myth and Thought among the Greeks*, trans. Janet Lloyd (New York: Zone Books, 2006), 375.

3. Michel Serres, *The Birth of Physics*, trans. Jack Hawkes (Manchester: Clinamen Press, 2000).

4. Bernard Williams, *Shame and Necessity* (Berkeley and Los Angeles: University of California Press, 1993).

5. Antonio Damasio, *Descartes' Error* (New York: Avon Books, 1994). For a more recent analysis that takes into account the revolution in neuroscience occasioned by the discovery of "mirror neurons," see Antonio and Hanna Damasio, "Minding the Body," *Daedalus* 135 (Summer 2006): 15–23.

6. Steven Shaviro, *Without Criteria: Kant, Whitehead, Deleuze and Aesthetics* (Cambridge, MA: MIT Press, 2009), 13.

7. Jacqueline de Romilly, *Time in Greek Tragedy* (Ithaca, NY: Cornell University Press, 1968), 88.

8. Augustine, *Concerning the City of God: Against the Pagans*, trans. Henry Bettenson (Harmondsworth, UK: Penguin, 1984), bk. 12, chap. 23, p. 503.

9. *The Confessions of St. Augustine*, trans. John K. Ryan (New York: Image Books, 1984), bk. 8, chap. 9, p. 197.

10. *St Augustine: Select Letters*, trans. James Baxter (Cambridge, MA: Harvard University Press, 1930), Epistle 179, p. 315.

11. Augustine, *On Christian Doctrine*, trans. D. W. Robertson (New York: Macmillan, 1958), 30.

12. Immanuel Kant, *Critique of Practical Reason*, trans. Lewis White Beck (New York: Library of Liberal Arts, 1993), 48–49.

13. Immanuel Kant, *Religion within the Limits of Reason Alone*, trans. Theodore H. Greene (New York: Harper, 1960), 32.

14. Kant, "On the Proverb: 'That May Be True in Theory But Is of No Practical Use,'" in *Perpetual Peace and Other Essays*, trans. Ted Humphrey (Indianapolis: Hackett, 1983), 86.

15. Kant, *Religion within the Limits of Reason Alone*, 110.

16. Jürgen Habermas, *The Future of Human Nature* (Cambridge, UK: Polity Press, 2003), 115.

17. I explore new developments in complexity theory in relation to the work of Alfred North Whitehead in chapter 1 of *A World of Becoming* (Durham, NC: Duke University Press, 2010). Indeed, I think of this essay as a companion to themes developed in that book.

18. All three statements come from *Critique of Practical Reason*, 153. For a book that explores the role that humiliation plays in the crystallization of practical reason and the willingness to submit to it, see Paul Saurette, *The Kantian Imperative: Humiliation, Common Sense, Politics* (Toronto: University of Toronto Press, 2005).

19. Benjamin Libet, Anthony Freeman, and Keith Sutherland, eds., *The Volitional Brain: Towards a Neuroscience of Free Will* (New York: Imprint Academic, 1999).

20. I have discussed elsewhere *how* to proceed, particularly in *Neuropolitics: Thinking, Culture, Speed* (Minneapolis: University of Minnesota Press, 2002) and *Pluralism* (Durham, NC: Duke University Press, 2005). For a book that carries such explorations to a higher pitch yet, see Stephen White, *The Ethos of a Late Modern Citizen* (Cambridge, MA: Harvard University Press, 2009).

21. Friedrich Nietzsche, *Twilight of the Idols* trans. R. J. Hollingdale (New York: Penguin, 1968), 73. See also "At Noon," in *Thus Spoke Zarathustra*, trans. Walter Kaufmann (New York: Penguin, 1978).

22. See *The Oedipus Plays of Sophocles*, trans. Paul Roche (New York: Penguin Books, 1996). The Roche translation is valuable because of its attention to the rhythms of the text and its attempt to capture in English some of the multiple meanings informing key speeches.

5

FREUD'S HELPLESSNESS

1. Michael Taylor in the introduction to the Penguin Classics *Richard III* (London: Penguin Books, 2005), xxv.

2. Samuel Beckett, *The Poems, Short Fiction, and Criticism of Samuel Beckett* (New York: Grove Press, 2006), 451.

3. Leo Bersani, *The Culture of Redemption* (Cambridge, MA: Harvard University Press, 1990).

4. *The Standard Edition of the Complete Psychological Works of Sigmund Freud* (London: Hogarth Press, 1856–1939) (hereafter, *SE*), 14:313.

5. Ibid.

6. *SE* 14:315.

7. Ibid.

8. *Psychoanalysis in France*, ed. Serge Lebovici and Daniel Widlöcher (New York: International Universities Press, 1980), 187.

9. Charles Taylor, *Sources of the Self: The Making of the Modern Identity* (Cambridge, MA: Harvard University Press, 1989), 3.

10. *SE* 1:379–81.

11. Alasdair C. MacIntyre, *Dependent Rational Animals: Why Human Beings Need the Virtues* (Chicago: Open Court, 1999), 5.

12. *The Origins of Psychoanalysis: Letters, Drafts and Notes to Wilhelm Fliess, 1887–1902* (Garden City, NY: Doubleday & Company, 1957), 379–80.

13. *SE* 21:30.

14. *SE* 11:123.

15. *SE* 20:154–155.

16. Taylor, *Sources of the Self*, 257.

17. Ibid., 215.

18. Friedrich Nietzsche, *On the Genealogy of Morals* (Oxford: Oxford University Press, 1996), 173.

19. Ibid., 20, 19, 9.

20. Dennis Donoghue, *On Eloquence* (New Haven: Yale University Press, 2008), 164.

21. *The Uncanny*, The New Penguin Freud (London: Penguin Books, 2003), 33.

22. Hilary Putnam, *Jewish Philosophy as a Guide to Life: Rosenzweig, Buber, Lévinas, Wittgenstein* (Bloomington: Indiana University Press, 2008), 79.

6

A SECULAR WONDER

I am grateful to Jonathan Davies, Stefanie Knauss, George Levine, and Davide Zordan for valuable comments on earlier versions of this essay.

1. Friedrich Nietzsche, *The Birth of Tragedy and Other Writings*, trans. Ronald Speirs (Cambridge: Cambridge University Press, 1999), 22–23. For an erudite discussion of the old saying, see Umberto Curi, *Meglio non essere nati. La condizione umana tra Eschilo e Nietzsche* (Turin: Bollati Boringhieri, 2008), chap. 1. Cf. also David Benatar, *Better Never to Have Been: The Harm of Coming into Existence* (Oxford: Oxford University Press, 2006).

2. Giacomo Leopardi, *Zibaldone*, n. 4174, April 19, 1826; in English, *Zibaldone: A Selection*, trans. Martha King and Daniela Bini (New York: Peter Lang, 1992).

3. M. Weber, *Max Weber: A Biography*, trans. Harry Zohn (New Brunswick, NJ: Transaction Publishers, 1988), 678. See Jane Bennett, *The Enchantment of Modern Life: Attachments, Crossings, and Ethics* (Princeton, NJ: Princeton

University Press, 2001), for a full-fledged discussion of the "Disenchantment Tales" and their danger of producing an "enervating cynicism" (13).

4. See Charles Taylor, *A Secular Age* (Cambridge, MA: Harvard University Press, Belknap Press, 2007), chap. 1.

5. See Hannah Arendt, *The Human Condition* (Chicago: University of Chicago Press, 1958), chap. 5.

6. Martha Nussbaum, "Transcending Humanity," in *Love's Knowledge: Essays on Philosophy and Literature* (Oxford: Oxford University Press, 1990), 366; see also the discussion in Taylor, *A Secular Age*, 623–30. From an opposite point of view, a contemporary caustic novelist like Milan Kundera has more prosaically explained the Greek hero's choice as the product of an optical illusion caused by the malfunctioning of human memory (which is constantly deceived by the impossible dream of the "Great Return"). Accordingly, he drastically discards the idea that we should take it as a proof that earthly life, as the "return," to quote Homer again (*Odyssey*, book 11, lines 100 and 203), is "honey-sweet" (*meliedea*). See Milan Kundera, *Ignorance*, trans. Linda Asher (New York: HarperCollins, 2002), §§ 2 and 9.

7. Taylor, *A Secular Age*, 311.

8. On the issue, see Stephen Mulhall, "Can There Be an Epistemology of Moods?" in *"Verstehhen" and Humane Understanding*, ed. Anthony O'Hear (Cambridge: Cambridge University Press, 1996), 191–210.

9. Sigmund Freud, *Civilization and Its Discontents*, trans. James Strachey (New York: W. W. Norton & Co., 1962), 11. For a thorough analysis of the issue, see William B. Parsons, *The Enigma of the Oceanic Feeling: Revisioning the Psychoanalytic Theory of Mysticism* (Oxford: Oxford University Press, 1999).

10. Freud, *Civilization and Its Discontents*, 12. The sensation is similarly described by Herman Melville in a famous letter to Hawthorne (June 1851): "This 'all' feeling, though, there is some truth in. You must often have felt it, lying on the grass on a warm summer's day. Your legs seem to send out shoots into the earth. Your hair feels like leaves upon your head. This is the all feeling," quoted in Richard Ruland and Malcolm Bradbury, *From Puritanism to Postmodernism: A History of American Literature* (London: Routledge, 1991), 143.

11. On this very influential epistemological conception, see Charles Taylor, "Foundationalism and the Inner-Outer Distinction," in *Reading McDowell: On Mind and World*, ed. Nicholas H. Smith (London: Routledge, 2002), 106–19. See also the now classic Gilbert Ryle, *The Concept of Mind* (Harmondsworth: Penguin, 2000).

12. Freud, *Civilization and Its Discontents*, 12–13.

13. Edmund Burke, *A Philosophical Enquiry into the Origin of Our Ideas of the Sublime and the Beautiful* (Oxford: Oxford University Press, 2008), 53. See also Stephen Greenblatt, "Resonance and Wonder," in *Exhibiting Cultures*, ed. Ivan Karp and Steven D. Lavine (Washington, DC: Smithsonian Institution Press, 1991), 49: "Looking may be called enchanted when the act of attention draws a circle around itself from which everything but the object is excluded, when intensity of regard blocks out all circumambient images, stills all murmuring voices"; and, by the same author, *Marvelous Possessions: The Wonder of the New World* (Oxford: Clarendon Press, 1991), 20: "The object that arouses wonder is so new

the following passage in
ıs. Alden L. Fisher (Pitts-
ıll the movements of the
s, one can, if one wishes,
ı the same way, since all
ırn been possible only by
ıosing the receptor to the
ır is the first cause of the
ıy the organism itself, by
tside. Doubtless, in order
of physical and chemical
—according to the proper
es and the movements of
world to which it will be

ʁting pure sensations (for
ʁhat they, as it were, "re-
ıe, *A Philosophical Testa-*
134.
d to moods, see Ryle, *The*
underline that we do not
nce the different kinds of

les Guignon, "Moods in
*Classic Readings in Phil-*
ʁrt C. Solomon (Oxford:
yfus, *Being-in-the-World:*
ision I* (Cambridge, MA:

ıds and Objectless Emo-
ıo. 1 (2006): 49–68, at
ʒ_price.pdf.
nsidered," in *Heidegger,*
ʁert L. Dreyfus,* ed. Mark
IT Press, 2000), 268. For
motions and moods, see
*Moral Psychology* (Cam-

ʁisco: North Point Press,
mology of Moods?" 202.
*ıl Review* 102 (1997):, 20.
word, see Peter L. Berger,
*Experience* (Berlin: Wal-

24. Francis Bacon, *The Advancement of Learning and Novum Organum* (New York: Colonial Press, 1899), 5, quoted in Bynum, "Wonder," 5 n. 16.

25. John Onians, "'I Wonder . . .': A Short History of Amazement," in *Sight and Insight: Essays on Art and Culture in Honour of E. H. Gombrich at 85*, ed. John Onians (London: Phaidon Press, 1994), 11.

26. Kelly Bulkeley, *The Wondering Brain* (New York: Routledge, 2005), 3. Along the same line, see also Jeff Malpas, "Beginning in Wonder: Placing the Origin of Thinking," in *Philosophical Romanticism*, ed. Nikolas Kompridis (London: Routledge, 2006), 284–85: "wonder [is] a response to the often sudden and striking *encounter* with things."

27. René Descartes, *The Passions of the Soul*, trans. Stephen H. Voss (Indianapolis: Hackett, 1989), 56 (art. 70).

28. John Benson, *Environmental Ethics: An Introduction with Readings* (London: Routledge, 2000), 73. See also Howard L. Parsons, "A Philosophy of Wonder," *Philosophy and Phenomenological Research* 30 (1969–1970): 85–86, and Malpas, "Beginning in Wonder," 284–89.

29. Albertus Magnus, *Metaphysica*, 1.2.6, quoted in James V. Cunningham, *Woe or Wonder: The Passional Effect of Shakespearean Tragedy* (Denver, CO: University of Denver Press, 1951), 77.

30. See Ronald Hepburn, "Wonder," in *"Wonder" and Other Essays* (Edinburgh: Edinburgh University Press, 1984), 131–54. See also Martha Nussbaum, *Upheavals of Thought: The Intelligence of Emotions* (Cambridge: Cambridge University Press: Cambridge, 2001), 54 n. 53: "Wonder is outward-moving, exuberant. . . . In wonder I want to leap or run."

31. Descartes, *The Passions of the Soul*, 52 (art. 53). Following a path already trodden by Hannah Arendt, Klaus Held has also emphasized the connection between wonder and birth; see Klaus Held, "Fundamental Moods and Heidegger's Critique of Contemporary Culture," trans. Anthony J. Steinbock, in *Reading Heidegger: Commemorations*, ed. John Sallis (Bloomington: Indiana University Press, 1993), 286–303. On the role of wonder and interest of the world even in the first infancy, see Nussbaum, *Upheavals of Thought*, 189–90.

32. See Lorraine Daston and Katharine Park, *Wonders and the Order of Nature: 1150–1750* (New York: Zone Books, 1998), chap. 8.

33. See Jenefer Robinson, "Startle," *Journal of Philosophy* 92, no. 2 (1995): 53–74.

34. Ibid., 57: "[startle] is a developmentally early form of two emotions in particular, namely, fear and surprise."

35. Ibid., 61.

36. Ibid., 62. See also 72: "Speaking very broadly—indeed, metaphorically—we can say that when we are startled, our bodies 'judge' that we are in a potentially threatening situation that needs immediate attention, but we must remember that this 'judgment' does not require cognitive activity or the formulation of any beliefs or conception of the situation in propositional form." It is interesting, but not surprising, that a similar functional explanation can also be applied to moods; see, e.g., Price, *Affect without Object*, 59ff.

37. Charles Darwin, *Notebook M*, 108 (August 26, 1838), in *Charles Darwin's Notebooks, 1836–1844: Geology, Transmutation of Species, Metaphysical*

that for a moment at least it is alone, unsystematized, *an utterly detached object of rapt attention*" (italics mine).

14. Freud, *Civilization and Its Discontents*, 12.

15. For an illuminating summary of this view, see the following passage in Marcel Merleau-Ponty, *The Structure of Behavior*, trans. Alden L. Fisher (Pittsburgh: Duquesne University Press, 1963), 13: "Since all the movements of the organism are always conditioned by external influences, one can, if one wishes, readily treat behavior as an effect of the milieu. But in the same way, since all the stimulations which the organism receives have in turn been possible only by its preceding movements which have culminated in exposing the receptor to the external influences, one could also say that the behavior is the first cause of the stimulations. Thus the form of the excitant is created by the organism itself, by the proper manner of offering itself to actions from outside. Doubtless, in order to be able to subsist, it must encounter a certain number of physical and chemical agents in its surroundings. But it is the organism itself—according to the proper nature of its receptors, the thresholds of its nerve centres and the movements of the organs—which chooses the stimuli in the physical world to which it will be sensitive."

16. One might add that the reason our few and fleeting pure sensations (for example, of pleasure or pain) are so overwhelming is that they, as it were, "replace" the whole world. On this point, see Marjorie Grene, *A Philosophical Testament* (Chicago: Open Court, 1995), chap. 7, especially 134.

17. For a similar use of the weather image with regard to moods, see Ryle, *The Concept of Mind*, 96, 100. It is, however, important to underline that we do not only suffer moods, but we *live* them, just as we experience the different kinds of weather (grudgingly, gleefully, miserably).

18. For a broader discussion of the topic, see Charles Guignon, "Moods in Heidegger's *Being and Time*," in *What Is an Emotion? Classic Readings in Philosophical Psychology*, ed. Cheshire Calhoun and Robert C. Solomon (Oxford: Oxford University Press, 1984), 230–43; Hubert L. Dreyfus, *Being-in-the-World: A Commentary on Heidegger's "Being and Time," Division I* (Cambridge, MA: MIT Press, 1991), chap. 10.

19. See Carolyn Price, "Affect without Object: Moods and Objectless Emotions," *European Journal of Analytic Philosophy* 2, no. 1 (2006): 49–68, at https://www.ffri.hr/phil/casopis/content/volume_2/eujap3_price.pdf.

20. See George Downing, "Emotion Theory Reconsidered," in *Heidegger, Coping, and Cognitive Science: Essays in Honor of Hubert L. Dreyfus*, ed. Mark A. Wrathall and Jeff Malpas, vol. 2 (Cambridge, MA: MIT Press, 2000), 268. For an interesting discussion of the relationship between emotions and moods, see also Robert C. Roberts, *Emotions: An Essay in Aid of Moral Psychology* (Cambridge: Cambridge University Press, 2003), 112–15.

21. Stanley Cavell, *The Senses of Walden* (San Francisco: North Point Press, 1981), 125, quoted in Mulhall, "Can There Be an Epistemology of Moods?" 202.

22. Caroline W. Bynum, "Wonder," *American Historical Review* 102 (1997):, 20.

23. For an elucidation of this untranslatable German word, see Peter L. Berger, *Redeeming Laughter: The Comic Dimension of Human Experience* (Berlin: Walter de Gruyter & Co., 1997), 35.

24. Francis Bacon, *The Advancement of Learning and Novum Organum* (New York: Colonial Press, 1899), 5, quoted in Bynum, "Wonder," 5 n. 16.

25. John Onians, "'I Wonder . . .': A Short History of Amazement," in *Sight and Insight: Essays on Art and Culture in Honour of E. H. Gombrich at 85*, ed. John Onians (London: Phaidon Press, 1994), 11.

26. Kelly Bulkeley, *The Wondering Brain* (New York: Routledge, 2005), 3. Along the same line, see also Jeff Malpas, "Beginning in Wonder: Placing the Origin of Thinking," in *Philosophical Romanticism*, ed. Nikolas Kompridis (London: Routledge, 2006), 284–85: "wonder [is] a response to the often sudden and striking *encounter* with things."

27. René Descartes, *The Passions of the Soul*, trans. Stephen H. Voss (Indianapolis: Hackett, 1989), 56 (art. 70).

28. John Benson, *Environmental Ethics: An Introduction with Readings* (London: Routledge, 2000), 73. See also Howard L. Parsons, "A Philosophy of Wonder," *Philosophy and Phenomenological Research* 30 (1969–1970): 85–86, and Malpas, "Beginning in Wonder," 284–89.

29. Albertus Magnus, *Metaphysica*, 1.2.6, quoted in James V. Cunningham, *Woe or Wonder: The Passional Effect of Shakespearean Tragedy* (Denver, CO: University of Denver Press, 1951), 77.

30. See Ronald Hepburn, "Wonder," in *"Wonder" and Other Essays* (Edinburgh: Edinburgh University Press, 1984), 131–54. See also Martha Nussbaum, *Upheavals of Thought: The Intelligence of Emotions* (Cambridge: Cambridge University Press: Cambridge, 2001), 54 n. 53: "Wonder is outward-moving, exuberant. . . . In wonder I want to leap or run."

31. Descartes, *The Passions of the Soul*, 52 (art. 53). Following a path already trodden by Hannah Arendt, Klaus Held has also emphasized the connection between wonder and birth; see Klaus Held, "Fundamental Moods and Heidegger's Critique of Contemporary Culture," trans. Anthony J. Steinbock, in *Reading Heidegger: Commemorations*, ed. John Sallis (Bloomington: Indiana University Press, 1993), 286–303. On the role of wonder and interest of the world even in the first infancy, see Nussbaum, *Upheavals of Thought*, 189–90.

32. See Lorraine Daston and Katharine Park, *Wonders and the Order of Nature: 1150–1750* (New York: Zone Books, 1998), chap. 8.

33. See Jenefer Robinson, "Startle," *Journal of Philosophy* 92, no. 2 (1995): 53–74.

34. Ibid., 57: "[startle] is a developmentally early form of two emotions in particular, namely, fear and surprise."

35. Ibid., 61.

36. Ibid., 62. See also 72: "Speaking very broadly—indeed, metaphorically—we can say that when we are startled, our bodies 'judge' that we are in a potentially threatening situation that needs immediate attention, but we must remember that this 'judgment' does not require cognitive activity or the formulation of any beliefs or conception of the situation in propositional form." It is interesting, but not surprising, that a similar functional explanation can also be applied to moods; see, e.g., Price, *Affect without Object*, 59ff.

37. Charles Darwin, *Notebook M*, 108 (August 26, 1838), in *Charles Darwin's Notebooks, 1836–1844: Geology, Transmutation of Species, Metaphysical*

*Enquiries*, ed. P. H. Barrett et al. (Cambridge: Cambridge University Press, 1987), 546 (italics mine).

38. Charles Darwin, *The Expression of the Emotions in Man and Animals* (Oxford: Oxford University Press, 1998), 278, 280.

39. For a convincing dispositional understanding of emotions, see Charles Taylor, "Explaining Action," *Inquiry* 13 (1970): 66–75.

40. See Daston and Park, *Wonders*, 303–16.

41. Ibid., 123.

42. Martin Heidegger, *Being and Time*, trans. John Macquarrie and Edward Robinson (Oxford: Blackwell, 2000), § 36, p. 216. On wonder and curiosity, see also Martin Heidegger, *Basic Questions in Philosophy: Selected "Problems" of "Logic"*, trans. Richard Rojcewicz and André Schuwer (Bloomington: Indiana University Press, 1994), §§ 37–38.

43. Burke, *A Philosophical Enquiry*, 29.

44. Heidegger, *Being and Time*, 217.

45. Daston and Park, *Wonders*, 307.

46. Ibid., 310.

47. See Hans Blumenberg, *The Legitimacy of the Modern Age*, trans. Robert M. Wallace (Cambridge, MA: MIT Press, 1983), 229–456.

48. Charles Darwin, *The Voyage of the Beagle* (London: Murray, 1860), 280 (italics mine).

49. Malpas, "Beginning in Wonder," 287 (italics mine).

50. A good example of this reverberation or, so to say, spreading of wonder "ripples" is the recent flood of personal memories elicited by the fortieth anniversary of the moon landing. In this case, the reaction to a specific amazing feat (man's first steps on the moon) imperceptibly but constantly turns into a broader wonder-response to much more ordinary biographical events connected to the big one.

51. Czeslaw Milosz, *Vision from San Francisco Bay*, trans. Richard Lourie (New York: Farrar, Straus and Giroux, 1982), 3.

52. See Cora Diamond, "The Difficulty of Reality and the Difficulty of Philosophy," in Stanley Cavell, Cora Diamond, John McDowell, Ian Hacking, and Cary Wolfe, *Philosophy and Animal Life* (New York: Columbia University Press, 2008), 43–89. By "difficulty of reality" she means "experiences in which we take something in reality to be resistant to our thinking it, or possibly to be painful in its inexplicability, difficult in that way, or perhaps awesome and astonishing in its inexplicability. *We take things so*. And the things we take so may simply not, to others, present the kind of difficulty, of being hard or impossible or agonizing to get one's mind around" (45–46). See also Greenblatt, *Marvelous Possessions*, 20–21: "The expression of wonder stands for all that cannot be understood, that can scarcely be believed . . . at once unbelievable and true."

53. Diamond, "The Difficulty of Reality," 61.

54. Ibid., 44. Cora Diamond is commenting here on the celebrated last stanza of Ted Hughes's poem "Six Young Men": "That man's not more alive whom you confront / And shake by the hand, see hale, hear speak loud, / Than any of those six celluloid smiles are, / Nor prehistoric or fabulous beast more dead; / No thought so vivid as their smoking blood: / To regard this photograph might well

dement, / Such contradictory horrors here / Smile from the single exposure and shoulder out / One's own body from its instant and heat."

55. I distinguish the just-mentioned five fundamental or basic moods from more derivative, reactive, or culturally determined *Stimmungen* (such as irritability, fastidiousness, contentiousness, peacefulness, naughtiness, insecurity, self-assurance, awkwardness, etc.).

56. While I interpret Diamond's "difficult of reality" in terms of the world's "superabundance" or "hyperdensity" of meaning, I do not exclude in the least the possibility that it may be personally lived as senseless, as when we are forced to come to terms with something emotionally unbearable. We are confronted here with two different meanings of "meaninglessness." After all, an event—for example, the untimely death of a beloved one—can be astonishingly senseless *just because* it fails to make sense in light of the thick, tangled web of meanings we experience in reality; because that tangled web is *too* fraught with meanings. "Superabundance" must not be equated with consistency.

57. For a persuasive defense of this idea, see Bennett, *The Enchantment of Modern Life*.

58. On humor, see Berger, *Redeeming Laughter*; Simon Critchley, *On Humour* (London: Routledge, 2002); Marie Collins Swabey, *Comic Laughter: A Philosophical Essay* (New Haven: Yale University Press, 1961); Helmuth Plessner, *Laughing and Crying: A Study of the Limits of Human Behavior*, trans. James Spencer Churchill and Marjorie Grene (Evanston, IL: Northwestern University Press, 1970).

59. See Berger, *Redeeming Laughter*, chap. 1.

60. Ibid., 61.

61. Immanuel Kant, *Critique of Judgment*, trans. Werner S. Pluhar (Indianapolis: Hackett, 1987), § 54, 203–5.

62. See Critchley, *On Humour*, 50–52.

63 Berger, *Redeeming Laughter*, 202. For a sensible interpretation of moderns' ambivalent attitude toward the comical, see also 215.

64. See Taylor, *A Secular Age*.

65. On the inescapable dialectic between opacity and transparency in our way of being-in-the-world, see Malpas, "Beginning in Wonder", 291–93.

66. Pico della Mirandola, "On the Dignity of Man," trans. Elizabeth L. Forbes, in *The Renaissance Philosophy of Man*, ed. E. Cassirer, P. O. Kristeller, and J. H. Randall Jr. (Chicago: University of Chicago Press, 1948), 224–25.

67. See Hannah Arendt, "Charlie Chaplin: Der Suspekte," in *Die verborgene Tradition. Acht Essays* (Frankfurt a.M.: Suhrkamp, 1976), 61.

## 7
### Prehuman Foundations of Morality

1. F. S. Collins, *The Language of God: A Scientist Presents Evidence for Belief* (New York: Free Press, 2006).

2. Sharpton made this assertion in a debate at the New York Public Library, on May 7, 2007, viewable at http://fora.tv.

3. F.B.M. de Waal, "Putting the Altruism Back into Altruism: The Evolution of Empathy, *Annual Review of Psychology* 59 (2008): 279–300.

4. C. Zahn-Waxler, B. Hollenbeck, and M. Radke-Yarrow, M., "The Origins of Empathy and Altruism," in *Advances in Animal Welfare Science*, ed. M. W. Fox and L. D. Mickley (Washington, DC: Humane Society of the United States, 1984), 21–39.

5. E. Hatfield, J. T. Cacioppo, and R. L. Rapson, *Emotional Contagion* (Cambridge: Cambridge University Press, 1994).

6. N. Eisenberg. "Empathy and Sympathy," in *Handbook of Emotion*, ed. M. Lewis and J. M. Haviland-Jones (New York: Guilford Press, 2000), 677–91.

7. M. L. Hoffman, "Is Altruism Part of Human Nature?" *Journal of Personality and Social Psychology* 40 (1981): 121–37.

8. S. D. Preston and F.B.M. de Waal, "Empathy: Its Ultimate and Proximate Bases," *Behavioral and Brain Sciences* 25 (2002): 1–72.

9. R. M. Yerkes, *Almost Human* (New York: Century, 1925), 246.

10. N. N. Ladygina-Kohts, *Infant Chimpanzee and Human Child: A Classic 1935 Comparative Study of Ape Emotions and Intelligence* (1935), ed. F.B.M. de Waal (New York: Oxford University Press, 2001), 121.

11. F.B.M. de Waal, *Good Natured: The Origins of Right and Wrong in Humans and Other Animals* (Cambridge, MA: Harvard University Press, 1996), and *Bonobo: The Forgotten Ape* (Berkeley and Los Angeles: University of California Press, 1997).

12. S. M. O'Connell, "Empathy in Chimpanzees: Evidence for Theory of Mind?" *Primates* 36 (1995): 397–410.

13. F.B.M. de Waal and A. van Roosmalen, "Reconciliation and Consolation among Chimpanzees," *Behavioral Ecology and Sociobiology* 5 (1979): 55–66.

14. F.B.M. de Waal and F. Aureli, "Consolation, Reconciliation, and a Possible Cognitive Differences between Macaque and Chimpanzee," in *Reaching into Thought: The Mind of the Great Apes*, ed. A. E. Russon, K. A. Bard, and S. T. Parker (Cambridge: Cambridge University Press, 1996), 80–110.

15. O. Fraser, D. Stahl, and A. Aureli, "Stress Reduction through Consolation in Chimpanzees," *Proceedings of the National Academy of Sciences, USA* 105 (2008): 8557–62.

16. E.g., J. Kagan, "Human Morality Is Distinctive," *Journal of Consciousness Studies* 7 (2000): 46–48; E. Fehr and U. Fischbacher, "The Nature of Human Altruism," *Nature* 425 (2003): 785–91.

17. De Waal, "Putting the Altruism Back into Altruism."

18. C. D. Batson, *The Altruism Question: Toward a Social-Psychological Answer* (Hillsdale, NJ: Erlbaum, 1991).

19. F.B.M. de Waal, *Chimpanzee Politics: Power and Sex among Apes* (1982) (Baltimore, MD: Johns Hopkins University Press, 2007), 67.

20. A. H. Harcourt and F.B.M. de Waal, *Coalitions and Alliances in Humans and Other Animals* (Oxford: Oxford University Press, 1992).

21. F. Warneken, B. Hare, A. P. Melis, D. Hanus, and M. Tomasello, "Spontaneous Altruism by Chimpanzees and Young Children," *PLoS Biology* 5 (2007): e184.

22. J. M. Burkart, E. Fehr, C. Efferson, and C. P. van Schaik, "Other-Regarding Preferences in a Non-human Primate: Common Marmosets Provision Food Altruistically," *Proceedings of the National Academy of Sciences* 104 (2007): 19762–66.

23. De Waal, F. B. M., Leingruber, K and Greenberg, A. R. (2008), "Giving is Rewarding for Monkeys," *Proceedings of the National Academy of Sciences, USA* 105:13685-13689.

24. W. T. Harbaugh, U. Mayr, and D. R. Burghart, "Neural Responses to Taxation and Voluntary Giving Reveal Motives for Charitable Donations," *Science* 326 (2007): 1622–25.

25. R. L. Trivers, "The Evolution of Reciprocal Altruism," *Quarterly Review of Biology* 46 (1971): 35–57.

26. F.B.M. de Waal and M. L. Berger, "Payment for Labour in Monkeys," *Nature* 404 (2000): 563.

27. Trivers, "The Evolution of Reciprocal Altruism"; F.B.M. de Waal, "The Chimpanzee's Service Economy: Food for Grooming," *Evolution of Human Behavior* 18 (1997): 375–86.

28. E. Fehr and K. M. Schmidt, "A Theory of Fairness, Competition, and Cooperation," *Quarterly Journal of Economics* 114 (1999): 817–68.

29. De Waal, *Good Natured*, 95.

30. S. F. Brosnan and F.B.M. de Waal, "Monkeys Reject Unequal Pay," *Nature* 425 (2003): 297–99.

31. M. van Wolkenten, S. F. Brosnan, and F.B.M. de Waal, "Inequity Responses of Monkeys Modified by Effort," *Proceedings of the National Academy of Sciences, USA* 104 (2007): 18854–59.

32. S. Freud, *Totem and Taboo: Some Points of Agreement between the Mental Lives of Savages and Neurotics* (1913) (New York: Norton, 1950).

33. See F.B.M. de Waal, *Primates and Philosophers: How Morality Evolved* (Princeton, NJ: Princeton University Press, 2006), for the appropriate distinctions and debate about what is unique about human morality.

8

## THE TRUTH IS SACRED

1. D. S. Wilson, "Species of Thought: A Comment on Evolutionary Epistemology," *Biology and Philosophy* 5 (1990): 37–62; *Darwin's Cathedral: Evolution, Religion, and the Nature of Society* (Chicago: University of Chicago Press, 2002); and "Rational and Irrational Beliefs from and Evolutionary Perspective," in *Rational and Irrational Beliefs*, ed. D. David, S. J. Lynn, and A. Ellis (Oxford: Oxford University Press, 2009).

2. B. Malinowski, *Magic, Science and Religion and Other Essays* (Boston: Beacon, 1948).

3. E. Durkheim, *The Elementary Forms of Religious Life* (1912), trans. K. E. Fields (New York: The Free Press, 1995).

4. Wilson, *Darwin's Cathedral*; "Testing Major Evolutionary Hypotheses about Religion with a Random Sample," *Human Nature* 16 (2005): 382–409; "Evolution and Religion: The Transformation of the Obvious," in *The Evolution*

*of Religion: Studies, Theories, Critiques*, ed. J. Bulbulia, R. Sosis, E. Harris, R. Genet, C. Genet, and K. Wyman (Santa Margarita, CA: Collins Foundation Press, 2008), 11–18; and "The Golden Rules of Religion," in *The Golden Rule: The Ethics of Reciprocity in World Religions*, ed. J. Neusner and B. D. Chilton (New York: Continuum, 2009).

5. E. Hutchins and G. E. Hinton, "Why the Islands Move," *Perception* (1984): 13, 629–32; E. Hutchins, *Cognition in the Wild* (Boston: MIT Press, 1996).

6. D. S. Wilson, "Language as a Community of Interacting Belief Systems: A Case Study Involving Conduct toward Self and Others," *Biology and Philosophy* 10 (1995): 77–97; and *Evolution for Everyone: How Darwin's Theory Can Change the Way We Think about Our Lives* (New York: Delacorte, 2007).

7. L. Margulis, *Origin of Eukaryotic Cells* (New Haven: Yale University Press, 1970).

8. J. Maynard Smith and E. Szathmary, *The Major Transitions of Life* (New York: W. H. Freeman, 1995); and *The Origins of Life: From the Birth of Life to the Origin of Language* (Oxford: Oxford University Press, 1999).

9. M. Tomasello, J. Carpenter, J. Call, T. Behne, and H. Moll, "Understanding and Sharing Intentions: The Origins of Cultural Cognition," *Behavioral and Brain Sciences* 28 (2005): 675–735; and M. Tomasello, *Why We Cooperate* (Boston: MIT Press, 2009).

10. A. Ehrenpreis, "An Epistle on brotherly community as the highest command of love" (1650), in *Brotherly Community: The Highest Command of Love*, ed. R. Friedmann (Rifton, NY: Plough Publishing Co., 1978), 66–69; see D. S. Wilson and E. Sober, "Reintroducing Group Selection to the Human Behavioral Sciences," *Behavioral and Brain Sciences* 17 (1994): 585–654, for additional discussion of this passage from an evolutionary perspective.

11. E.g., S. Brown and E. Dissanayake, "The Arts Are More than Aesthetics: Neuroaesthetics as Narrow Aesthetics," in *Neuroaesthetics*, ed. M. Skov and O. Vartanian (Amityville, NY: Baywood, 2009), 43–57.

12. Personal communication.

13. E.g., S. Pinker, *The Blank Slate: The Modern Denial of Human Nature* (New York: Viking, 2002).

14. E.g., G. F. Miller, *The Mating Mind: How Sexual Choice Shaped the Evolution of Human Nature* (New York: Doubleday, 2000).

15. M. Horton, *The Long Haul* (New York: Doubleday, 1990), 7.

16. J. Queenan, *Closing Time: A Memoir* (New York: Viking, 2009).

17. R. H. Thaler and C. R. Sunstein, *Nudge: Improving Decisions about Health, Wealth, and Happiness* (New Haven: Yale University Press, 2008).

18. T. Berry, *The Great Work: Our Way into the Future* (New York: Harmony, 1999).

19. C. Boehm, *Hierarchy in the Forest: Egalitarianism and the Evolution of Human Altruism* (Cambridge, MA: Harvard University Press, 1999); R. Dunbar, C. Knight, and C. Power, eds., *The Evolution of Culture* (New Brunswick, NJ: Rutgers University Press, 1999).

20. http://theevolutioninstitute .org

21. http://evolution.binghamton.edu/bnp

22. Hutchins, *Cognition in the Wild*.

23. S. C. Hayes, "Acceptance and Commitment Therapy, Relational Frame Theory and the Third Wave of Behavioral and Cognitive Therapies," *Behavior Therapy* 35 (2004): 639–65.

24. M. Dowd, *Thank God for Evolution: How the Marriage of Science and Religion Will Transform Your Life and Our World* (New York: Plume, 2009).

25. http://www.babasword.com/index/rgecd.html

## 9
### DARWINIAN ENCHANTMENT

1. The spirit world has not yet ceased to be. / Your mind is closed, your heart is dead. / Come now, you novice, bathe sans misery / Your earthly breast in morning's hue of red.

2. Max Weber, "Wissenschaft als Beruf," in Max Weber, *Schriften 1894–1922*, ed. Dirk Kaesler (Stuttgart: Kröner, 2002), 498.

3. Ibid., 488.

4. George Bernard Shaw, "Preface," *Back to Methuselah* (1921) (London: Penguin Books, 1961), 44.

5. William Whewell, *History of the Inductive Sciences, from Earliest to Present Times* (1837), 3rd ed., 3 vols. (London: Parker & Son, 1857). The changes in the subsequent editions are marked in separate sections. The main text is that of the first edition of 1837.

6. Ibid., 3:485

7. Ibid., 478.

8. Ibid., 476.

9. Darwin obtained a copy of Whewell's volumes shortly after they were published in 1837. See Charles Darwin to Charles Babbage (June–September 1837), in *The Correspondence of Charles Darwin*, ed. Frederick Burkhardt et al., 17 vols. to date (Cambridge: Cambridge University Press, 1985–), 2:22.

10. Whewell, *History of the Inductive Sciences*, 3:480.

11. Charles Darwin, *Origin of Species* (London: Murray, 1859), 490.

12. Whewell, *History of the Inductive Sciences*, 3:489.

13. Charles Darwin, *E Notebook*, 60, in *Charles Darwin's Notebooks, 1836–1844*, ed. Paul Barrett et al. (Ithaca: Cornell University Press, 1987), 414.

14. Darwin, *C Notebook*, 63, in *Darwin's Notebooks*, 259.

15. Charles Darwin, *The Autobiography of Charles Darwin*, ed. Nora Barlow (London: Collins, 1958), 120.

16. In summer of 1838, Darwin jotted in his *D Notebook:* "The Varieties of the domesticated animals must be most complicated, because they are partly local & then the local ones are taken to fresh country & breed confined to certain best individuals.—scarcely any breed but what some individuals are picked out.—in a really natural breed, not one is picked out."

17. For a complementary conclusion, see Robert J. Richards, "Darwin's Theory of Natural Selection and Its Moral Purpose," in *Cambridge Companion to Darwin's Origin of Species*, ed. Michael Ruse and Robert J. Richards (Cambridge: Cambridge University Press, 2008), 47–66.

18. Darwin, *D Notebook*, 134e–135e, in *Darwin's Notebooks*, 375.

19. There are those, like Michael Ghiselin, who believe Darwin's theory left no room at all for teleology. See his *The Triumph of the Darwinian Method* (Berkeley and Los Angeles: University of California Press, 1969), 132–59. Others, like Michael Ruse, recognize what might be called "internal teleology," that is, that traits of organisms serve a function and have been selected for that function. Ruse, though, maintains that Darwin did not assume a designer or the operations of mind: "he [Darwin] showed how to get purpose without directly invoking a designer—natural selection gets things done according to blind law without making direct mention of mind. The teleology is internal." See Michael Ruse, *Darwin and Design: Does Evolution Have a Purpose?* (Cambridge, MA: Harvard University Press, 2003), 126.

20. Darwin, *B Notebook*, 49, in *Darwin's Notebooks*, 182.

21. Darwin, *E Notebook*, 48–49, in *Darwin's Notebooks*, 409.

22. In addition to the passages already cited, the following employ explicit consideration of "final causes": Darwin, *B Notebook*, 5, in *Darwin's Notebooks*, 171; *C Notebook*, 236, in *Darwin's Notebooks*, 313; *D Notebook*, 114e, in *Darwin's Notebooks*, 369; *E Notebook*, 146 and 147, in *Darwin's Notebooks*, 440.

23. Darwin, *M Notebook*, 154, in *Darwin's Notebooks*, 559.

24. Charles Darwin, *The Descent of Man and Selection in relation to Sex*, 2 vols. (London: Murray, 1871), 1:136.

25. Darwin, *Autobiography*, 87. Darwin was mistaken about Paley, who also thought the Creator worked through natural laws.

26. Darwin, *M Notebook*, 36–37, in *Darwin's Notebooks*, 343.

27. William Paley, *Natural Theology* (London: Faulder, 1809), 416.

28. William Whewell, *Astronomy and General Physics Considered with Reference to Natural Theology* (Bridgewater Treatise) (Philadelphia: Carey, Lea & Blanchard, 1833), 267.

29. Charles Darwin, *Charles Darwin's natural Selection, being the Second Part of his Big species Book written from 1856 to 1858*, ed. R. C. Stauffer (Cambridge: Cambridge University Press, 1975), 224.

30. Charles Darwin to Asa Gray (May 22, 1860), in *Correspondence of Charles Darwin*, 8:224.

31. These essays have been transcribed by Francis Darwin, and published in *The Foundations of the Origin of Species: Two Essays Written in 1842 and 1844 by Charles Darwin* (Cambridge: Cambridge University Press, 1909).

32. Darwin, essay of 1842, in *Foundations of the Origin of Species*, 5. In the *Origin of Species*, Darwin assumes that small individual variations in progeny will be constant. However, even in the book he occasionally suggests that favorable variations will occur only sporadically over great periods of time (e.g., 80 and 82).

33. Darwin, essay of 1842, 8.

34. Ibid., 5.

35. Ibid., 85.

36. Darwin, *Origin of Species*, 83–84.

37. Ibid., 62–63.

38. Ibid., 41.

39. Ibid., 105. Darwin frequently reiterated the need for large numbers of individuals for selection to work most efficiently; see 41, 70, 110, 125, 177, and 179.

40. Ibid., 84.

41. Ibid.

42. Ibid., 489.

43. In addition to the passages mentioned, see also ibid., 83, 149, 194, and 201.

44. Ibid., 78.

45. Ibid., 84.

46. Darwin, *E Notebook*, 49, in *Darwin's Notebooks*, 409; see above for the full quotation.

47. Darwin began his *N Notebook* on October 2, 1838, and completed it in spring of 1839. He also kept a series of loose notes on morality and mind, which he labeled "Old & Useless Notes," which date roughly from summer of 1838 through early 1840.

48. Darwin, *Old & Useless Notes*, 39, in *Darwin's Notebooks*, 616.

49. Darwin, *Old & Useless Notes*, 37, in *Darwin's Notebooks*, 614.

50. Darwin, essay of 1844, in *Foundation of the Origin of Species*, 254. This line, of course, occurs with slight alteration both in the essay of 1842 and in the *Origin of Species*.

51. In the *Origin of Species* (236), Darwin mentioned that the problem initially seemed "fatal to my whole theory." I have discussed this problem in greater detail in my *Darwin and the Emergence of Evolutionary Theories of Mind and Behavior* (Chicago: University of Chicago Press, 1987), 142–52.

52. Darwin, *Descent of Man*, 1:166.

53. Ibid., 163.

54. Ibid., 97–98.

55. Ibid., 100–101.

56. I have suggested a few amendments to Darwin's moral theory to defend it against many of the usual objections brought against evolutionary ethics. See, for example, appendix 2 of my *Darwin and the Emergence of Evolutionary Theories of Mind and Behavior*. For an empirical investigation of the notion of a "moral grammar," see Marc Hauser, *Moral Minds* (Cambridge, MA: Harvard University Press, 2005); see as well John Mikhail, "Universal Moral Grammar: Theory, Evidence and the Future," *Trends in Cognitive Sciences* 11 (2007): 143–52. For further empirical confirmation that individuals do act with authentic altruism, see Ernst Fehr and Urs Fischbacher, "The Economics of Strong Reciprocity," in *Moral Sentiments and Material Interests*, ed. Herbert Gintis et al. (Cambridge, MA: MIT Press, 2005), 151–92.

57. Aveling, the consort of Karl Marx's daughter, published this account in a little tract just after Darwin's death. He quotes Darwin as saying: "I never gave up Christianity until I was forty years of age," which would have been around 1850. See E. B. Aveling, *The Religious Views of Charles Darwin* (London: Freethought Publishing Co., 1883), 5. Francis Darwin confirmed Aveling's account in Francis Darwin, ed., *Life and Letters of Charles Darwin*, 3 vols. (London: Murray, 1887), 1:317n.

58. Darwin, *Autobiography*, 92–93.

59. Alfred Russel Wallace to Darwin (July 2, 1866), in *Correspondence of Charles Darwin*, 14:228.

60. See Darwin to Alfred Russel Wallace (July 5, 1866, in ibid., 236.

61. Darwin, *Origin of Species*, 489.

62. There is every good reason to believe, as well, that a world comparable to our own can be found as the productive result of the interactions of the higher animals in the world.

63. James had recruited the concept of natural selection to discuss the activities of consciousness in decision making. See my discussion in *Darwin and the Emergence of Evolutionary Theories of Mind and Behavior,* 430–35. For Donald Campbell's use of the concept, see his "Blind Variation and Selective Retention in Creative Thought as in Other Knowledge Processes," *Psychological Review* 67 (1960): 380–400; and "Blind Variation and Selective Retention in Socio-Cultural Evolution," in *Social Change in Developing Areas*, ed. H. Barringer, G. Blanksten, and R. Mack (Cambridge, MA: Schenkman, 1965).

## 10
### The Wetfooted Understory: Darwinian Immersions

1. Salman Rushdie, "Is Nothing Sacred?" in *Imaginary Homelands: Essays and Criticism 1981–1991* (London: Granta, 1991), 415–16.

2. Ibid., 423.

3. Ibid., 420–21.

4. See Leo Bersani's *The Culture of Redemption* (Cambridge, MA: Harvard University Press, 1992) for further discussion of this.

5. See Gillian Beer, *Darwin's Plots: Evolutionary Narrative in Darwin, George Eliot and Nineteenth-Century Fiction* (Cambridge: Cambridge Univesity Press, 1983).

6. For a particularly eloquent example, read Don Paterson's essay on his "conversion to scientific materialism" in his extraordinary "version" of Rilke's *Die Sonette an Orpheus*, entitled *Orpheus: A Version of Rilke* (London: Faber, 2006), 65–68.

7. Darwin, *The Autobiography of Charles Darwin*, ed. Nora Barlow (London: Collins, 1958), 138–39.

8. See, for instance, a sermon preached by John Piper, pastor for preaching at Bethlehem Baptist Church, Minnesota, called "Learn from Darwin: We Become What We Behold" at www.desiringgod.org.

9. Darwin, *Autobiography*, 136.

10. Ibid., 21.

11. Harold Bloom, *The Anxiety of Influence* (Oxford: Oxford University Press, 1974).

12. See Toby Appel, *The Cuvier-Geoffroy Debate: French Biology in the Decades before Darwin* (Oxford: Oxford University Press, 1987); Martin Rudwick, *Bursting the Limits of Time: The Reconstruction of Geohistory in the Age of Revolution* (Chicago: Chicago University Press, 2007); and Derinda Outram,

*George Cuvier: Vocation, Science and Authority in Post-Revolutionary France* (Manchester: Manchester University Press, 1984).

13. Coleridge to Wordsworth, May 30, 1815, *Coleridge's Collected Letters* (Oxford: Clarendon Press, 2002), 4:574–75.

14. See the chapter called "On Speculating" in my book *Darwin and the Barnacle* (London: Faber, 2003), 135–53.

15. Ernst von Hesse-Wartegg, "Bei Charles Darwin" [At Charles Darwin's], *Frankfurter Zeitung und Handelsblatt* (July, 30, 1880): 1–2. [English translation—Darwin Correspondence Project.]

16. Adam Phillips, *Darwin's Worms* (London: Faber, 1999), 55.

17. Darwin, *The formation of vegetable mould, through the action of worms, with observations on their habits* (London: John Murray, 1881), 313.

18. Phillips, *Darwin's Worms*, 60–61.

19. W. Tegetmeier, "Darwin on Orchids," *Register of facts and occurrences relating to literature, the science, and the arts*, August 1862, 38–39.

20. Traubel, *With Walt Whitman in Camden*, ed. Sculley Bradley (Philadelphia: University of Pennsylvania Press, 1953), 4:392.

21. Ibid., 8:454.

22. Whitman, *Specimen Days and Collect*, from *Complete Poetry and Collected Prose* ed. Justin Kaplan, Literary Classics of the United States (New York: Library of America, 1982), 897–98.

23. Elizabeth Bishop letter to Anne Stevenson, January 8–20, 1964, Elizabeth Bishop Papers, Washington University, St. Louis.

24. Bishop to Robert Lowell, Letter 302, July 30, 1964, in *Words in Air: The Complete Correspondence between Elizabeth Bishop and Robert Lowell*, ed. Robert Travisano (London: Faber and Faber, 2008), 545.

25. Bishop to Robert Lowell, Letter 303, July 30, 1964, in *Words in Air*, 547.

26. Bishop to Robert Lowell, Letter 382, October 14, 1971, in *Words in Air*, 692.

27. Lowell to Bishop, Letter 292, April 6, 1964, in *Words in Air*, 528.

28. Bishop to James Merrill, June 3, 1971, in *One Art: Selected Letters* (London: Pimlico, 1996), 542.

29. Darwin to Hooker, June 13, 1850, in *The Correspondence of Charles Darwin*, ed. Frederick Burkhardt et al., 17 vols. to date (Cambridge: Cambridge University Press, 1985–), 4:344.

30. Darwin, *Notebook M*, 34–35, http://darwin-online.org.uk.

31. By the mid-nineteenth century the word "speculation" was being used in a derogatory sense to mean both (in science) "abstract or hypothetical reasoning on subjects of a deep, abstruse, or conjectural nature" and (in the financial world) "engagement in any business enterprise or transaction of a venturesome or risky nature, but offering the chance of great or unusual gain."

32. Darwin, *The Cross and Self-Fertilisation of Plants* (London: John Murray, 1876), 458.

33. Ernesto Suarez-Toste, "Une Machine a Coudre Manuelle: Elizabeth Bishop's 'Everyday Surrealism,'" *Mosaic* (Winnipeg), June 1, 2000.

34. Elizabeth Bishop letter to Anne Stevenson, January 8–20, 1964, Elizabeth Bishop Papers, Washington University, St. Louis.

35. Guy Rotella, *Reading and Writing Nature: The Poetry of Robert Frost, Wallace Stevens, Marianne Moore, and Elizabeth Bishop* (Boston: Northeastern University Press, 1991), 188.

36. Ibid., 189–90.

37. Darwin, *Autobiography*, 91–92.

38. William Darwin, Note, Cambridge University Library, DAR112.B3b—B3f page sequence 6–7.

39. George Darwin and Francis Darwin, eds., "Darwin celebration, Cambridge, June, 1909. Speeches delivered at the banquet held on June 23rd," *Cambridge Daily News*, June 24, 1909, 15.

40. According to Hesse-Wartegg's account of 1880, Darwin still had Drosera growing in the house twenty years later. Hesse-Wartegg, "Bei Charles Darwin."

# Index